POTENTIAL USE OF SOLAR ENERGY AND EMERGING TECHNOLOGIES IN MICRO IRRIGATION

POTENTIAL USE OF SOLAR ENERGY AND EMERGING TECHNOLOGIES IN MICRO IRRIGATION

Editor

Jeltsyn Rocha

scitus
academics

Potential Use of Solar Energy and Emerging Technologies in Micro Irrigation

Edited by **Jeltsyn Rocha**

ISBN: 978-1-68117-708-3
Library of Congress Control Number: 2016958745

Notice

Printed in United States of America on Acid Free Paper

PREFACE

Farmers have always played an important role in our society as they provide food to the world's population. Nevertheless, one could forget that not only is food supplied by them, it also supplies energy, which is actually absolutely necessary today, especially in view of renewable energies. Indeed, farmers can create power from the wind, biomass or the sun and can use it for their own farm or perhaps resell it to companies if they have a surplus. Access to the electricity grid is not guaranteed in many rural areas, in particular in developing and emerging countries. Farmers cannot rely on the traditional irrigation system in this particular circumstance. Consequently, using an alternative and independent energy system could be a solution for the farmer to secure a safe source of energy and to keep the public grid from saturation alone. Solar energy could be one of the easiest ways to produce energy for farmers. In fact, farmers usually have many large buildings whose roofs are right under the sun, without being hampered by the shadows of the trees, making them a great place to set up a solar system. The use of solar energy in agriculture is therefore becoming increasingly popular, and the power generated from this renewable source can be used either on the farm or even in the regional energy grid if the farmer has an extra income.

This book examines the possibility of solar energy and other evolving micro-irrigation energy technologies to create sustainable energy solutions. It discusses several innovative designs of micro-irrigation systems with a unique focus on solar power and photovoltaic (PV) energy. In addition, nutrient management is deliberated through gout irrigation in order to improve crop quality and yield. This book is a valuable tool for urban development professionals, energy professionals, graduate students, as well as for practitioners and researchers. R&D today tends to focus on the production of solar pumps that are affordable in dry regions. As a result, the gout irrigation system could contribute to efficient water management in countries with high and rare water resources. This is even more important because farmers are facing three challenges: saving water, energy and cash. Movable solar gout irrigation systems are the best solution for these challenges. Although these methods are still very expensive and complex to solve, many R&D projects are working hard to democratize the use of solar energy in agriculture, which could play a brilliant role in the management of food and energy emergencies in the future and even today.

TABLE OF CONTENTS

1

KEY IRRIGATION TECHNOLOGIES AND SUBSTRATE CHOICE FOR SOILLESS POTTED FLOWERS IN GREENHOUSES

Fusheng Ma[1] and Haiyan Fan[2]

[1] Beijing Hydraulic Research Institute, Beijing, China

[2] Beijing Engineering Technique Research Center for Exploration and Utilization of Non-Conventional Water Resources and Water Use Efficiency, Beijing, China

1. INTRODUCTION

Total area of flower production in China has reached 834, 000 hm^2 In 2009, with the total sales of ¥719.80 billions; the planting area of potted flower has reached 81 710.6 hm^2, with the sales amount of ¥180.80 billions, the amount of protected cultivation area for flower has reached 81767.5 hm^2, where the area of greenhouse is 21 490.5 hm^2 (The website of state forestry administration, 2009). Facility substrate culture has been one of the production models pushed mainly for urban modern agriculture, so it is very significant in practice for the application of soilless culture technology to find an economic, high-yield and efficient culture substrate suitable for the locality. According to statistics, the typical irrigation quota of facility flower is 3 150~3 900 m^3/hm^2 (Beijing Water Science and Technology Institute, 2007) with high quality and clean water resource, but problems such as undefined water consumption amount, low utilization efficiency of irrigation water, lack of irrigation program, wasting of good water and fertilizer resource, unscientific application of water and fertilizer that may impact economic quality of flower are existing.

It is inevitably demand to uncover water consumption of plant for research about natural hydrological cycle, exploration of plant ecological functions and guidance of plant precision irrigation, and assurance of high yield and quality of commercialcrop, thereby this area is always the hot spot of research in various disciplines such as hydrology, meteorology, irrigation and water conservancy, ecotope and so on. At present, there are many findings in water consumption of typical crops (GuoShenghu et al., 2010; ShenXiaojun et al., 2007; Zheng Baoguo et al., 2010; Zhao Ying et al., 2011), fruit trees (Zhao Jinghua et al., 2010; Liu Hongguang et al., 2010; Abrisqueta J M, 2008) and landscape plants (GuHongzhong et al., 2006; Philips N et al., 1996; Sun Huizhen et al., 2004), the research efforts for facility flower water-saving irrigation is still weak, where the research achievement for water consumption concerned soilless culture facility flower is rarely seen, which falls behind rapid development of flower industry seriously.

Research about the irrigation technology for soilless culture flower still lags behind for the walking tube-well irrigation wasting manpower and water-fertilizer resource and then ruining the environment, while the drip irrigation system cannot fit for changing of the density of potted flowers, which hard to manage. But the ebb-and-flow system can recycle water and fertilizer resource as a new irrigation technology. In 1990s, researchers started systematic study on influence to plant growth irrigated by ebb-and-flow technology from fertilizer and substrate (Poole R T et al., 1992; Erin James et al., 2001), comparison of water utilization between various irrigation technologies in a greenhouse (Catherine A. Neal et al., 1992) algae control in an ebb and flow irrigation system (Chase R et al., 1993), some get the patents (Robert W et al.,1994). At the beginning of the 21st century, Chinese researchers started to introduce the ebb-and-flow technology (Ren Jianhua et al., 2004; Yang Tieshun et al., 2009), which settled in Ningxia (GaoYuting et al., 2010), in 2011, a special planting bed was developed (Zhang Xiaowen et al., 2011), in addition to the ebb-and-flow culture test for potted*Hydrangea macrophylla*(Zhang Li et al., 2011), and get many patents of irrigating equipment quickly.

But the ebb-and-flow technology has not been used massively yet for the limitation of its large initial investment, high requirements for the operation and management personnel's technical level and the production facilities after it is introduced to China (Ma Fusheng et al., 2012).The drip arrow technology is used in production of potted plant cultivated with solid substrate as soon as its appearance (Meng Qingguo et al., 2005), the research achievements concentrate mainly on water-fertilizer coupling effect of vegetable, comparison and selection of proper substrate and determination of the lower limit of irrigation (Pang Yun et al., 2006; Yue bin et al., 1998; Wang Ronglian et al., 2009; ZhongGangqiong et al., 2005). Peat soil becomes the main substrate for soilless substrate potted flower, because which can provide available mosisture, fertilizer, air and temperature supplying enviroment. Based on the research results, though various irrigation low limits are key for water utilization efficiency, growth and quality of potted flower (De Boodt M et al., 1983; Zhang Tiejun et al., 2010; Wang Yajun et al., 2003; Zhang Jianhua et al., 2009; Zhang Ning et al., 2011; R. KastenDumroese et al., 2011), there is seldom reports about research results of influence to water utilization efficiency, growth and quality of soilless substrate potted flower from various dripper flows. Scientific basis for selection and matching the drip irrigation emitters of soilless substrate potted flower are lack.

At present, there have been reports concerned influences to growth and development, yield and quality from soilless culture substrate by scholars both here and abroad (De Boodt M et al., 1983; Zhou Yanli et al., 2005; Zhang Tiejun et al., 2010; Wang Yanjun et al.,

2003; Zhang Jianhua et al., 2009; Zhang Ning et al., 2011; R. KastenDumroes et al., 2011; David W R. Water et al., 2006; Kang Hongmei et al., 2006; Zhao Jiuzhou et al., 2001; Li Jing et al., 2000; Chen Zhende et al., 1998; Tian Jilin et al., 2003). But the researches are only concentrated on raw material and suitability of substrate based on the indexes of plant growing, yield and quality; in addition to testing evaluation of the indexes such as pore structure, volume-weight and so on for substrate moisture physical property.

Based on the status quo of weak research efforts for high level facility flower water consumption and research lag on new irrigation technology, we take the typical potted foliage flower "Anthurium" as the object of study, the water consumption law and utilization efficiency during the overall process of the facility systematically were studied, in order to provide scientific basis for design of the irrigation system and precision irrigation control of facility potted anthurium. We take drip irrigation as a comparison to study influence of various nutrient solution depths to substrate moisture content, irrigation water utilization, plant growth and water consumption in order to provide scientific basis for the new technology of facility anthurium ebb-and-flow irrigation. For the sake of a proper dripper flow, we choose peat soil (PINDSTURP) that is popular in the market as the culture substrate, study the law of influence to water utilization, growth and quality of soilless potted flower from various dripper flows systematically, in order to provide basis for selection of facility substrate potted flower drippers. At the same time, proceeding from hydrodynamic parameter testing and evaluation of substrate, 8 typical substrates of facility soilless culture were chosen to discuss the water binding capacity, storativity and water availability of them, in order to provide enough theoretical basis for substrate comparison and selection.

2. EXPERIMENTAL SITE AND METHODS

2.1. Site and Experiment Description

- Beijing Jidinglida Technology & Trade Co., Ltd.

We start test and research about influence to water utilization and growth of facility drip irritation soilless culture anthurium under various irrigation low limits, facility soilless potted anthurium ebb-and-flow technology, facility potted anthurium proper dripper flow selection at the intelligent multi-span greenhouse of Beijing Jidinglida Technology & Trade Co., Ltd. (39°20′, east longitude 114°20′, elevation 12m).The material of the greenhouse is double hollow polycarbonate sheet with the span of 8m, and the standard width of a room is 4m. There are the temperature and humidity sensors in the rooms, which drive the wetted pads and the fans automatically. The wetted pads and fans are fitted on the south and north walls in the room, in addition to the small fans to accelerate air circulation inside. When the wetted pads and fans are operated, at the same time the door and the skylight are closed, air circulation can be improved and suitable environment is guaranteed. The greenhouse is heated by the heating radiators to keep the temperature between 20°C~30°C. There are external sun louver and internal sunshading to avoid flower leaf burns.

- Central Station for Irrigation Experiment of Beijing

The soilless culture substrate moisture characteristic parameter test and research are performed at the soil physics laboratory of Central Station for Irrigation Experiment of Beijing, Yongledian Town, Tongzhou, Beijing (39°20′, East longitude 114°20′, Elevation 12m).

2.2. Experimental Methods

- Water consumption

Weigh the flowerpot everyday with a balance that has a precision of 0.01g. When irrigating, monitor seepage loss collected by the water collector pot by port; weight the pot before and after irrigation, and determine stage water consumption by the water quantity balance method. Weigh the flower sample every other 5~7 days during the test, check the results to guarantee accuracy. Count the water consumption by the water consumption of single pot, and determine the stage water consumption with the water quantity balance method, refer to formula (1).

$$ET_i = W_1 - W_2 + I - D \tag{1}$$

Where, *ETi* is the water consumption at period I; *W1* is the flowerpot weight at the beginning of the period, g. Pot^{-1}; *W2* is the flowerpot weight at the end of the period, g. Pot^{-1}; *I* is the flowerpot irrigation during the period, g. Pot^{-1}; and *D* is the flowerport water leakage during the period, g. Pot^{-1}.

- Crown diameter and Plant height

Measure the crown diameter of the sample in the directions of east-west and north-south with a ruler, and take the mean values in both directions as the representative values of the sample crown diameter. Take the pot edge as the datum, measure the distance from the pot edge to the peak of the sample as the representative value of the plant height. Take the distance between the spathe to the root of the petiole as the length, and the distance of the widest as the width.

- Substrate moisture physical property

Measure the dry volume mass and the substrate water holding capacity of the cultural medium with the conventional methods (Jiang Shengde et al., 2006).

- Substrate moisture content

The dry substrate mass is determined by the dry unit weight filling before testing, the empty pot mass is determined by weighing before testing, and the overall mass of the flower and the pot is the average mass of the 5 sample pots. Take 3 plants from the same treatment in protecting line, which are same as the sample plant weighted, weigh their net mass and take the mean value as the representative value of plant mass, the testing period is about 5 days. Take the testing day as the middle, and 2 days both before and after it to use the mass value of the same plant, and calculate the moisture content of the substrate day by day. Subtract the plant mass, the dry substrate mass and the pot mass from the overall mass of the testing flower and the pot to get the water mass in the substrate. Divide the mass with water by the unit weight of water, and then by the substrate volume to get the moisture content of substrate.

- Substrate porosity

The saturated gravity drainage method is used to determine the porosity (Jiang Shengde et al., 2006), which including determination of total porosity, the water holding porosity, the aeration porosity and the gas water ratio. Weigh a vessel with a known volume (V) and weigh its mass (W_1), fill the testing substrate according to the design unit weight and weigh its mass (W_2). Soak the vessel with substrate for 24 hours and weigh its mass (W_3), encase the upper port of the vessel by wet gauze with a known mass (W_4) to avoid seepage of fine particle. There shall be no water seepage when inversing the vessel, and weigh its mass (W_5). The formula for the porosity of the substrate is:

$$\varphi = \left[W_3 - W_1 - \left(W_2 - W_1\right)\right] / V \times 100\% \tag{2}$$

$$\varphi_t = \left[W_3 + W_4 - W_5 \right] / V \times 100\% \tag{3}$$

$$\varphi_c = \left[W_5 - W_2 - W_4 \right] / V \times 100\% \tag{4}$$

$$d = \varphi_c / \varphi_t \tag{5}$$

Where φφ = Total porosity, %; φtφt = Aeration porosity, %; φcφc = Water holding porosity, %; d = Gas water ratio; the units of $W_1 \sim W_5$ are g; and the unit of V is cm^3.

- Substrate permeability coefficient

It is determined with the constant head method (Tian Jilin, 2003). Fill the cutting ring as required by the test, soak it for 24h before take off. Connect an empty cutting ring on it, seal the connecting place to avoid leakage. Add filter paper at the bottom of the connecting cutting ring. Add water into the empty cutting ring till the water level 1mm lower than the edge of the cutting ring. Time when the funnel starting dripping, record the seepages at 1, 3, 5, 7, 10, 15, 20min..., at the same time, add water into the empty cutting ring to the initial level and record the water temperature. End the test when the seepage becomes stable. Repeat each treatment for 3 times and average them. Hereinafter is the formula for the permeability coefficient:

The permeability coefficient at t°C

$$K_t = vL / \left(h + L \right) \tag{6}$$

Where vv — Permeability speed, mm min^{-1}; hh — Height of water layer, cm; LL — Height of substrate, cmThe permeability coefficient at 10°C

$$K_{10} = K_t / (0.7 + 0.03t) \tag{7}$$

- Substrate moisture characteristic curve

Determine the moisture contents of substrate under 0, 0.01, 0.03, 0.05, 0.1, 0.3, 0.5, 1 and1.5Mpa respectively with the 1500F1membrane pressure gauge and dray the moisture characteristic curve.

- Substrate horizontal diffusivity

Determine it with the horizontal soil columnInfiltration method (Lei Zhidong, 1988). Put the substrate (Or soil) with the air drying unit weight as required by the test into the organic glass tube with the diameter of 5cm and the length of 50cm, ignore the gravity, water it with the Markov bottle. Refer toFig. 1 for the test unit. Record the variation of water level in the bottle and the time when the wetting front passes every 1cm. At the end of the test, take soil (Or substrate) from the wetting front quickly, weigh it and dry it by heating, so get the moisture content distribution of the soil (Or substrate) column. With the testing time *t* and the moisture content distribution of the soil (Or substrate) column at that time, λλ corresponding to various θθ can be calculated with the formula

$\lambda = xt^{-1/2}$, so the formula for the horizontal diffusivity is:

$$D(\theta) = -\frac{1}{2}\frac{\Delta\lambda}{\Delta\theta}\sum_{\theta_0}^{\theta}\lambda\Delta\theta \tag{8}$$

Where $D(\theta)D(\theta)$ — Horizontal diffusivity, $cm^2\ min^{-1}$; $\theta 0\theta 0$ — Initial moisture content, $cm^3\ cm^{-3}$; $\theta\theta$ — Moisture content, $cm^3\ cm^{-3}$; $\lambda\lambda$ — Boltzmann transformation parameter, $\lambda = xt^{-1/2}$; x — , The moving distance of the wetting front at t, cm; t— Time, min

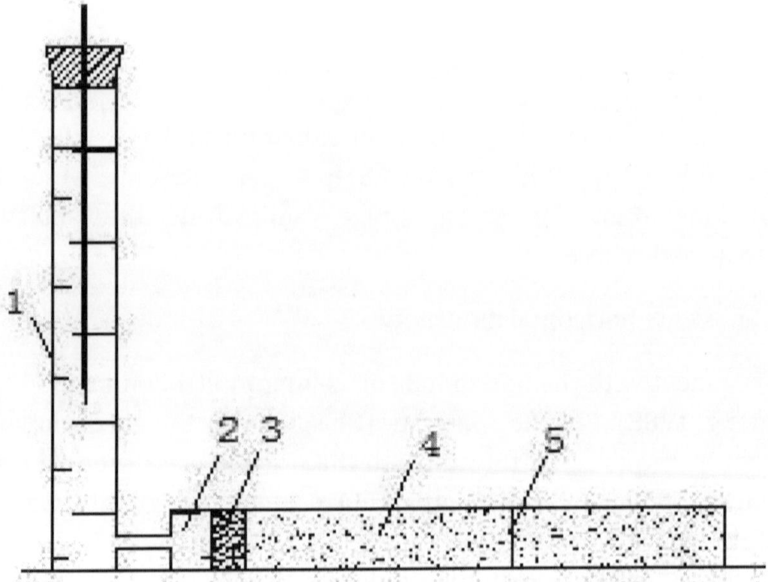

Figure 1. Experimental equipment of measuring soil water diffusivity

1. Markov bottle 2.Water chamber 3. Filter layer 4. Horizontal soil column 5. Wetting front

- Evaluation method

The matrix method is used for substrate comprehensive evaluation. A matrix method refers to form a matrix with all development activities and all environmental factors impacted, establish direct causality, in order to show the impact of the activity on the environmental factor (Duan Zhaolin, 2003). In this paper, the values in the order of merit of 8 substrates under influence from the total porosity, gas water ratio, permeability coefficient, water availability and transmissibility are formed, the better, the larger. Finally, perform comprehensive evaluation to various substrates according to the synthesis score.

3. IMPACT OF VARIOUS LOW IRRIGATION THRESHOLD ON WATER UTILIZATION AND GROWTH OF FACILITY DRIP SOILLESS CULTURE ANTHURIUM

3.1. Experimental Plant and Design

The anthurium is chosen as the testing object, its variety is "Flame", which belongs to *Araceae* and*Anthurium*. The culture substrate is mixture of PINDSTURP peat soil imported from Denmark and perlite from China in a volume ratio of 10:1, its dry volume mass is 0.16 g/cm^3, and the volume water holding rate (θFC) is 0.4315. The tested anthurium is cultivated in a pot, the top diameter of the pot is 16cm, the bottom diameter is 10cm, the height is 12.8cm, the volume is about 1.4L, and the substrate is filled about 2cm to the pot top. Refer to Table 1 for testing treatment.

Table 1. Experimentle treatments

Testing stage	Lower limit of irrigation/(%)				
	T1	T2	T3	T4	T5
Whole growing period	90	80	70	60	50

Note: The lower limit of irrigation is counted in percentage of the substrate water holding capacity.

In this paper, all treatment irrigation quota are controlled by the substrate water holding capacity as the upper limit. The test starts from July 4, 2009 when transplanting to November 5, 2009, takes 124d. The placing density of tested plantlet is 15 pot/m^{2}in the seeding stage, and 10 pot/m^2 in the later stage. The plantlet is 18cm in height with uniform growth when transplanting. Use the pressure compensating dripper with water yield of 3.85 L/h (The working pressure range is 0.05~3 MPa), with the flat perforated pipe configured 4 outlets. Connect the drip arrow with uniform 1m long capillary tube, insert shallowly in the pot substrate from the side direction about 1cm within the center area of the root system. Determine the irrigation quota with the substrate water holding rate as the upper limit according to the relative lower control limit. Place the pot on the bracket, and put the PVC collector under the bracket to collect water leakage. Repeat 5 times per treatment. There are the protection lines at both sides of the monitored sample plant. The self-prime stabilized pressure pump with the pressure tank is used; the hydraulicproportioning fertilization pump made by USA is used for fertilization with water.All test plots are distributed randomly, the treatment measures such as farming, fertilizing, and pest control are same.

3.2. Results and Discussion

3.2.1. Impact of various low irrigation threshold on moisture content of substrate with drip irrigation

Refer to Fig. 2 for influence to moisture content of substrate from various low irrigation threshold. At the beginning of the test, adjust the basic moisture contents of all treatment substrates to the water holding capacity of the substrate. There is no irrigation for all treatments from July 4 to 7, so the moisture contents of substrate of various treatments have no large difference as shown on Fig. 1. From July 7 to 12, T1, T2 and T3 reach the lower irrigation limit and start to irrigate, the moisture contents of the 3 treatment substrates go down in turn from T1 to T3, but all of them are higher than T4 and T5 that do not irrigate and have not very large difference in moisture contents. After July 20, the overall performance of the moisture contents of substrate are 82%~96%θ_{FC} of T1, 75%~94%θ_{FC} of T2, 64%~88%θ_{FC} of T3, 56%~80%θ_{FC} of T4 and 44%~76%θ_{FC} of T5, which goes down from T1 to T5 along with decreasing of the low irrigation threshold. It identifies with the research result of influence to the soil moisture content from the lower irrigation limit (Karam K. et al, 2011).

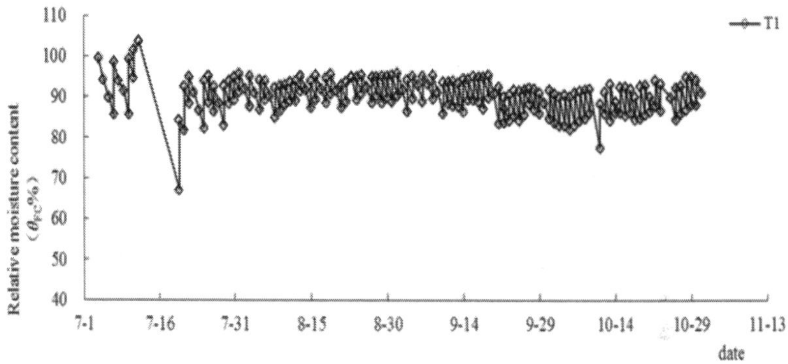

a. T1 (The lower irrigation limit is 90% of the water holding rate of substrate)

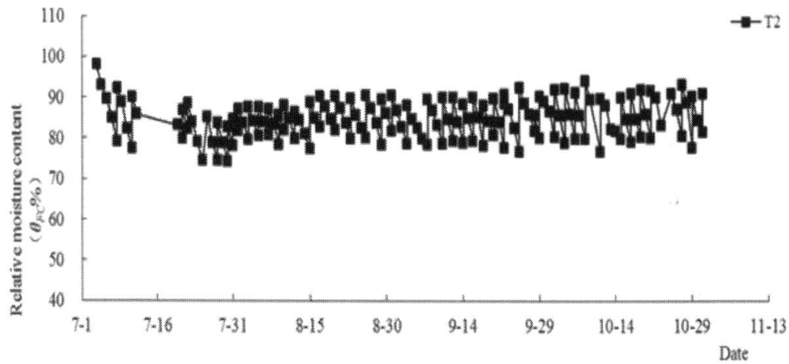

b. T2 (The lower irrigation limit is 80% of the water holding rate of substrate)

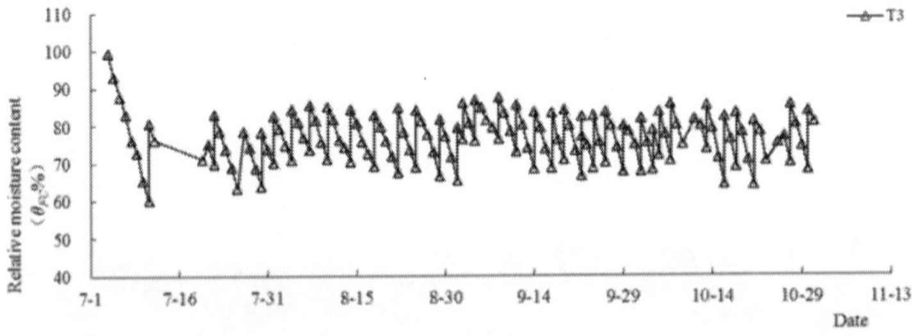

c. T3 (The lower irrigation limit is 70% of the water holding rate of substrate)

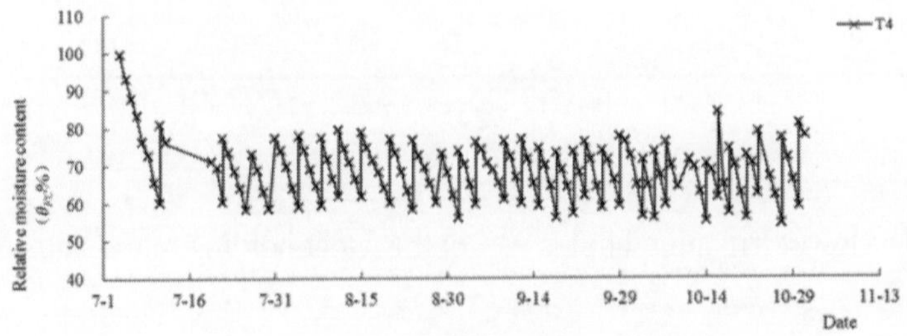

d. T4 (The lower irrigation limit is 60% of the water holding rate of substrate)

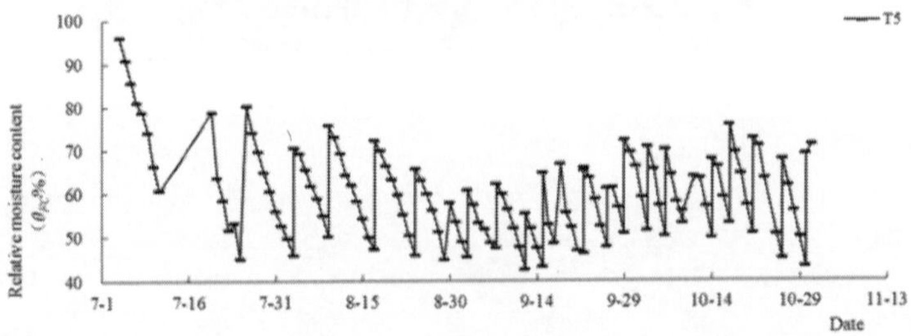

e. T5 (The lower irrigation limit is 50% of the water holding rate of substrate)

Note: The relative moisture contentis counted in percentage of the substrate water holding capacity.

Figure 2. The influence to moisture content of substrate from various low irrigation threshold

Refer to Fig. 3 for the influence to the substrate moister content from various low irrigation threshold after irrigation, while the influence to the field water utilization coefficient is shown in Fig. 4.

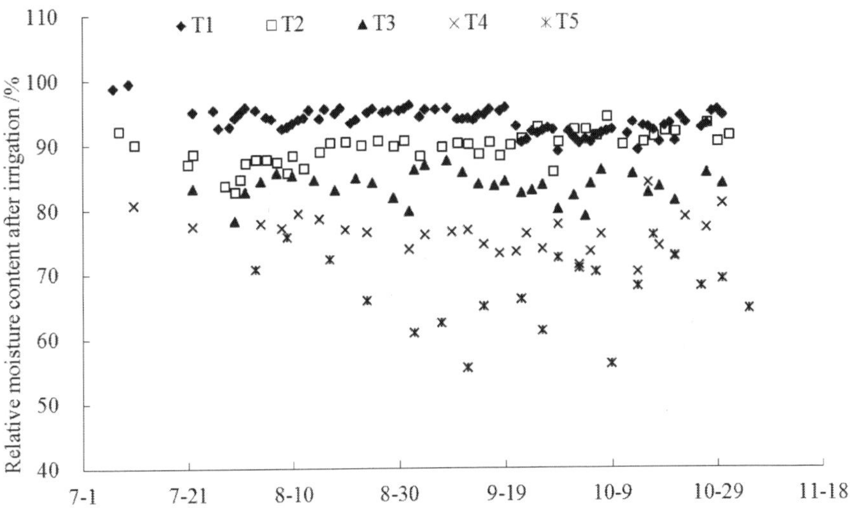

Note: The relative moisture content is counted in percentage of the substrate water holding rate.

Figure 3. The influence to the substrate moister content from various low irrigation threshold after irrigation

As shown in Fig. 3, the substrate moisture contents after irrigation go down along with decreasing of the low irrigation threshold. During the test, the mean values of T1, T2, T3, T4 and T5 after irrigation are 93.32%θ_{FC}, 89.31%θ_{FC}, 83.31%θ_{FC}, 76.02%θ_{FC} and 66.95%θ_{FC} respectively, all of them are lower than the level of substrate water holding capacity measured with conventional methods. It is caused by the substrate drip irrigation local water supply and the physical property of organic medium changing along with the substrate moisture content. Based on the measured results during the test, the average water leakage losses of all irrigations for

T1, T2, T3, T4 and T5 are 22.30, 49.05, 78.02, 107.74 and 138.62g/pot respectively, the irrigation leakage losses increase along with decreasing of the low irrigation threshold.

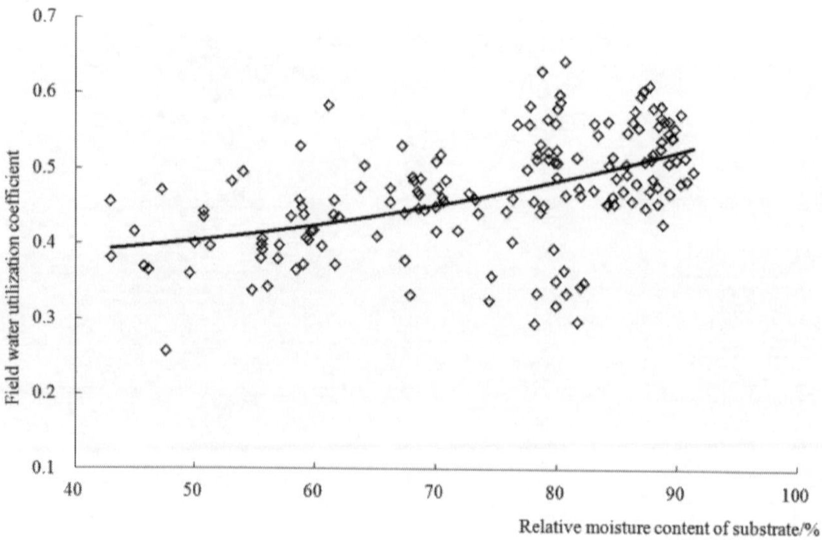

Note: The relative moisture content is counted in percentage of the substrate water holding rate.

Figure 4. Influence to the field water utilization coefficient from various low irrigation threshold after irrigation

The field water utilization coefficient is the ratio between the available water irrigated into field (For dry farmland, it refers to the irrigation water stored in the planned moisture layer) and water discharge from the last stage of the fixed ditch (Field ditch) (Guo Yuanyu et al., 2006). In this paper, it refers to the ratio between the water stored and absorbed by the substrate in the port and the total water irrigated in the port. The total water irrigated in the port can be measured, and the water stored by the substrate is the difference between the total water and leaked water. Refer to Fig. 4,under the test conditions, the total field water utilization coefficient goes down along with the substrate moisture content before irrigation. During the test, the average field water utilization coefficients of T1, T2, T3, T4 and T5 are 0.519, 0.471, 0.439, 0.419 and 0.402. While the irrigation water utilization coefficient specified in*Technical code for microirrigation engineering* (GB/T 50485-2009) shall not be less than 0.9, so the field water utilization coefficient in the tested substrate is even less than it.

In conclusion, the lower irrigation limit is the key factor for the substrate moisture and field irrigation water utilization coefficient (Both of them go down along with the lower irrigation limit) under the conditions that may affect water supply of drip irrigation. Based on their research about influence to substrate physicochemical properties from various water supply modes, Qi Haiying et al. (Qi Haiying et al., 2009) find that various water supply modes can affect the substrate pore structure, substrate volume, salinity, and acid and alkali environmentdramatically. So the change feature of the substrate moisture content and the field irrigation water utilization coefficient maybe the influence result to the tested culture substrate volume and pore structure from local water supply of the drip arrow, while the water absorption, storage and dissipation mechanism of soilless culture substrate needs more intensive study.

3.2.2. Influence to anthurium irrigation water amount from various low irrigation threshold treatments under drip irrigation conditions

Refer to Table 2 for influence to anthurium irrigation water capacity from various low irrigation threshold under drip irrigation conditions.

Table 2. Effect of different low irrigation threshold to irrigation resume of Anthurium

Treatment	Irrigation quota/ (g·pot⁻¹)	Irrigation period/d	Irrigation quota/ (g·pot⁻¹)
T1	46.36	All are 1~2 d during the test.	4865.4
T2	92.72	All are 2~3 d during the test.	5124.1
T3	139.08	3~4 d before 60 d after planting, and 2~3 d from 60 to 120d.	5953.9
T4	185.44	4~5 d before 60 d after planting, and 2~3 d from 60 to 120d.	6725.6
T5	231.80	5~7 d before 60 d after planting, and 3~5 d from 60 to 120d.	6344.0

As shown in Table 2, the irrigation period expands along with the decreasing of the lower irrigation limit, which is basically same as the conclusion of the routine soil water-saving irrigation test (Yang Wenbin, 2003; Tian Yi et al., 2006); but the irrigation quota of each treatment goes up along with decreasing of the lower irrigation limit. To compare with T1, the quota of T2, T3, T4 and T5 increases 5.3%, 22.7%, 38.2% and 29.3% respectively, which is on the contrary of the routine soil water-saving irrigation test (Sun Huayin et al., 2008). The field water utilization coefficient of the tested substrate goes down along with decreasing of the lower irrigation limit, it is the reason that causes the substrate culture irrigation quota increases along with decreasing of the lower irrigation limit.

3.2.3. Influence to anthurium growth status from various low irrigation threshold under drip conditions

Refer to Fig. 5a and Fig. 5b for influences to the plant height and crown diameter of the anthurium from various low irrigation threshold under drip conditions.

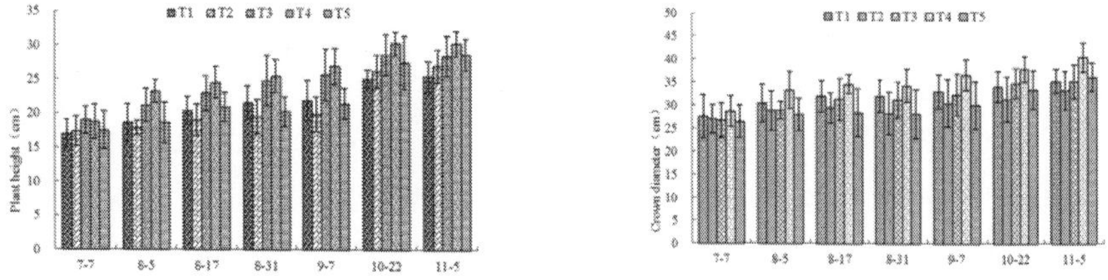

a Plant height b Crown diameter

Figure 5. Effect of different low irrigation threshold to plant height and crown diameter

As Fig. 5a shown, T3 and T4 appear obvious growth vigor. By the end of the test, the plant heights of T1, T2, T3, T4 and T5 are 25.4, 27.0, 28.5, 30.3 and 28.7 cm respectively. Based on the significance testing results, the height of T4 is 19.3% higher than T1 obviously; the differences between other treatments and T1 are not obvious.

As Fig. 5b shown, as with the plant height, the crown diameter of T4 appears obvious growth vigor. By the end of the test, the anthurium crowns of T1, T2, T3, T4 and T5 are 35.3, 33.3, 35.2, 40.5 and 36.2 cm respectively. Based on the significance testing results ($p=0.05$), the crown diameter of T4 is 14.7% higher than T1 obviously, but the differences between other treatments and T1 are not obvious.

Table 3. Effect of different low irrigation threshold to spathes

Treatment	Parameter			
	Qty. of spathe	Length of spathe/cm	Width of spathe/cm	Height of inflorescence/cm
T1	2.7±0.8	10.7±1.2	7.5±1.5	3.5±0.4
T2	2.4±0.5	10.4±2.3	7.3±1.5	3.4±1.0
T3	2.5±0.5	10.9±1.5	7.4±1.5	3.7±0.7
T4	2.6±0.4	10.9±1.9	7.8±1.3	4.3±0.7
T5	2.7±0.5	9.8±1.3	7.1±1.0	3.6±1.1

Under drip irrigation conditions, refer to Table 3 for influence to the ornamental part - the spathe of the anthurium from various low irrigation threshold. The flower quantity of the 5 treatments are all 3, there is no large difference among treatments. The lengths of the spathe of T1~T4 are all about 10.5cm, the widths are all about 7.5cm, and there is no large difference among treatments ($p=0.05$). The lowest lower irrigation limit – T5 has the smallest length and width of the spathe, it has negative influence to ornamental quality of the anthurium.

3.2.4. Influence to anthurium water consumption from various low irrigation threshold under drip irrigation conditions

Refer to Table 4 for influence to the anthurium water consumption from various low irrigation threshold under drip irrigation conditions.

Table 4. Effect of different low irrigation threshold to water consumption amount of Anthurium

Treatment	Anthurium water consumption of each treatment/(g·pot^{-1})				
	07-04~07-31	August	September	10-01~11-04	Total water consumption during experiment
T1	572.2	655.5	729.2	1028.9	2985.8
T2	554.4	559.5	599.1	849.3	2562.3
T3	554.4	632.2	696.0	922.4	2805.1
T4	578.8	661.8	699.3	918.4	2858.3
T5	557.3	534.2	538.6	891.7	2521.7

Based on Table 4, the water consumption of the tested anthurium during test is between 2 521~2 985 g/pot. To compare with T1, the total anthurium consumption of T2, T3, T4 and T5 are all reduced, in the degree of 14.2%, 6.1%, 4.3% and 15.5% respectively, viz. decreasing of the low irrigation threshold reduce the water consumption of the tested anthurium. Li Xia (Li Xia et al., 2010) carry out research about the transpirations of potted tomato with various substrate moisture contents, they also find that decreasing of the substrate moisture content reduces the transpiration of substrate potted tomato. Difference of water consumption between various treatments is caused by both the plant transpiration and substrate evaporation. Based on the plant heights (Fig. 5 (a)), growth of T1 and T2 is relatively weak; since the substrate moisture content of T1 is always at a higher level because of frequent irrigation, the main reason of large total water consumption is strong substrate evaporation; while the moisture content of T2 is lower than T1 obviously, the main reason of reduced water consumption may be weak substrate evaporation; on account of good growth of T3 and T4, their higher water consumption may be caused by strong transpiration; and the substrate moisture content of T5 is the lowest with weaker growth, substrate evaporation and plant transpiration, which may be the main reason to cause its lower water consumption. So water consumption mechanism of potted substrate needs intensive study.

3.2.5. Proper lower irrigation limit for soilless culture anthurium under drip irrigation conditions in a multi-span greenhouse

The characteristics of the flower industry are high input cost and seeking high returns, so the irrigation technology and the irrigation system are crucial to affect flower quality and benefit. Since the economic value of flower depends on quality, the high priority consideration when making the irrigation system shall be high quality of flower. At the same time, multiple benefits such as water saving, fertilizer saving, energy saving and manpower reducing shall be considered synthetically.

In this paper, the lower irrigation limit is key to water consumption and quality of the anthurium. To compare with T1, the flower count and the spathe of T4 that has the lower irrigation limit 60% of the substrate water holding capacity are not affected, but its water consumption has 4.3% reduction, the plant height has 19.3% increasing obviously, the crown diameter has 14.7% increasing obviously, which improves ornamental quality of the anthurium. While to compare T1 with others, though the water consumption goes down to varying degrees, there is no ornamental quality improvement, even negative effect to T5. So under the drip irrigation conditions in this paper, the proper lower irrigation limit of the anthurium shall be 60% of the substrate water holding capacity, here the substrate moisture content is between 60%~80% of the substrate water holding capacity; the irrigation period within 60d of transplanting is 4~5d, for 60~124d it is 2~3d, and the irrigation quota during the test is 185.44g/pot.

Though T4 has better irrigation effect in this paper, its field water utilization coefficient is only about 0.42, with serious leakage loss of good water and fertilizer resource from drip irrigation that become crucial environmental pollution source. It is urgently needed for intensive study on soilless culture substrate water absorption and storage mechanism, research and development of substrate culture drip irrigation technical mode, development of substrate with stable physicochemical property, reinforcement of circulation utilization of good water and fertilizer resource. Improve the utilization efficiency of water and fertilizer, meanwhile guaranteeing flower quality; support the flower industry, meanwhile achieve water-saving irrigation, good product quality and environmental protection and so on.

3.3. Conclusions

(1) The low irrigation threshold is the key controlling factor for the substrate moisture content and the irrigation water use efficiency. Both the substrate moisture content and the field water utilization coefficient go down along with the low irrigation threshold, while the irrigation leakage loss increases along with decreasing of the lower irrigation limit. Under the testing conditions in this paper, the field water utilization coefficient is only about 0.40~0.52, the irrigation period was prolonged along with the low irrigation

threshold, but the irrigation quota increases on the contrary. To compare with the treatment that has the prolong as much as 90% of the substrate water holding rate, the irrigation quotas of treatments that have the low irrigation threshold as much as 80%, 70%, 60% and 50% of the substrate water holding rate increase 5.3%, 22.7%, 38.2% and 29.3% respectively.

(2) Under the testing conditions, the difference of the anthurium flower count between various treatments is not large, all of them are 3. When the lower irrigation limit is 60% of the substrate water holding capacity, the spathe size of the anthurium is not affected. But the plant height and the crown diameter increase 19.3% and 14.7% ($p=0.05$) obviously, in addition to improving ornamental quality. For the other treatments, except for the treatment with the lower irrigation limit as much as 50% of the substrate water holding rate that has reduced anthurium spathe, the others have no obvious difference from the treatment with the lower irrigation limit as much as 90% of the substrate water holding rate.

(3) During the test, the anthurium water consumption is about 2 522~2 986 g/pot, and the water consumption goes down along with decreasing of the lower irrigation limit. To compare with the treatment with the lower irrigation limit as much as 90% of the substrate holding capacity, the water consumptions of the treatments with the lower irrigation limit as much as 80%, 70%, 60% and 50% of the substrate water holding rate reduce 14.2%, 6.1%, 4.3% and 15.5% respectively. In order to achieve the anthurium optimal ornamental quality under the test conditions, determine that the proper lower irrigation limit of the anthurium shall be 60% of the substrate water holding capacity. When the substrate moisture content is between 60%~80% of the substrate water holding capacity, the irrigation period within 60d of transplant is 4~5d, and it is 2~3d after 60d~124d, the irrigation quota is 185.44g/pot.

4. EBB-AND-FLOW IRRIGATION TECHNOLOGY FOR SOILLESS POTTED ANTHURIUM

4.1. Experimental Plant and Design

The test starts planting on July 4, 2009 and ends on November 5, takes totally 124d. The pot density to October 14 is 15 pot/m^2, and 10 pot/m^2 after that day. The planted seedlings have uniform growth, about 18cm high. The tested object is potted anthurium, its variety is "Flame", which belongs to*Araceae* and *Anthurium*. The culture substrate is mixture of PINDSTURP peat soil imported from Denmark and perlite from China in a volume ratio of 10:1, its dry volume mass is 0.16 g/cm^3, and the volume water holding rate (θFC) is 0.4315. For the tested pot, its top diameter is 16cm, its bottom diameter is 10cm, its height is 12.8cm, and its volume is about 1.4L. There are five treatments in the test, from T1 to T4 are ebb-and-flow irrigation, their nutrient solution depths are 1/2, 1/3, 1/4 and 1/5 of the pot height (H). T5 is drip irrigation treatment, as shown in Table 5.

Table 5. Experimental treatments

Growth stage	Add English version				
	T1	T2	T3	T4	T5
Whole growth period	Nutrient solution depth of $\frac{1}{2}H$	Nutrient solution depth of $\frac{1}{3}H$	Nutrient solution depth of $\frac{1}{4}H$	Nutrient solution depth of $\frac{1}{5}H$	Drip irrigation

Note: H is the pot height, here the pot height is 12.8cm.

The water channel bonding with transparent acrylic sheet is used as the culture vessel of ebb-and-flow irrigation. The tested pots are placed according to the production density. Control the liquid volume according to the set nutrient solution depth for each treatment, after 30min soak, discharge residual water. For drip irrigation treatment, use the pressure compensating dripper with water yield of 3.85 L/h (The working pressure range is 0.05~3 MPa), with the flat perforated pipe configured 4 outlets. Connect the drip arrow with uniform 1m long capillary tube, insert shallowly in the substrate within the center area of the flower root system. Place the pot on the bracket, and put the PVC collector under the bracket to collect water leakage;the lower irrigation limit of each treatment are unified as 80% of the substrate water holding rate (θ_{FC}),but the irrigation quota of drip irrigation takes the upper irrigation limit as 100%θFC. Repeat 5 times per treatment. There are the protective plants around the sample. The self-prime stabilized pressure pump with the pressure tank is used; the hydraulicproportioning fertilization pump made by USA is used for drip irrigation. But for ebb-and-flow irrigation, dissolve fertilizer in water directly for fertilization with water. Determine the times of drip irrigation and fertilization as required by production, the fertilization times of ebb-and-flow is same as drip irrigation. Distribute the 5 test areas randomly, management, protection, pest control, and fertilizer solution concentrations are same.

Table 6. Irrigation system for all treatments

Test treatment	Irrigation system
T1	The irrigation period for the whole growth period is 5~7 d, with an average of 5.6d. Irrigate nutrient solution till 1/2 of the pot height is reached.
T2	The irrigation period for the whole growth period is 4~6 d, with an average of 4.9d. Irrigate nutrient solution till 1/3 of the pot height is reached.
T3	The irrigation period for the whole growth period is 3~5 d, with an average of 3.7d. Irrigate nutrient solution till 1/4 of the pot height is reached.
T4	The irrigation period for the whole growth period is 3~5 d, with an average of 3.7d. Irrigate nutrient solution till 1/5 of the pot height is reached.
T5	The irrigation period for the whole growth period is 2~3 d, with an average of 2.6d. And the irrigation quota is 92.72g·pot^{-1}.

4.2. Results and Discussion

4.2.1. Dynamics of substrate moisture content

Refer to Fig. 6 for substrate moisture content after irrigation of each treatment. For all treatments, the substrate moisture content after irrigation appears slight fluctuation. The ranges of substrate moisture content after irrigation for T1, T2, T3, T4 and T5 are 106.5%~118.2%θFC, 96.7%~110.0%θFC, 90.2%~103.4%θFC, 89.4%~105.0%θFC and 82.7%~93.1%θFC respectively. The mean values of substrate moisture contents after irrigation are 112.8%θFC, 105.8%θFC, 95.9%θFC, 96.8%θFC and 89.1%θFC respectively. Based on those, the average relative substrate moisture content after ebb-and-flow irrigation is 6.8~23.7%θFC higher than that of drip irrigation. The moisture contents of T1 and T2 are higher than the other 3 treatments obviously. T1 is the highest, while T5 with drip irrigation treatment is the lowest; the moisture contents of T3 and T4 with ebb-and-flow irrigation are not very different. When the nutrient solution depth does not exceed 1/4 of the pot height, the influence from the nutrient solution depth to the relative substrate moisture rate after irrigation is not large; but when it exceeds 1/4 of the pot height, the influence increases obviously along with increasing of the nutrient solution depth.

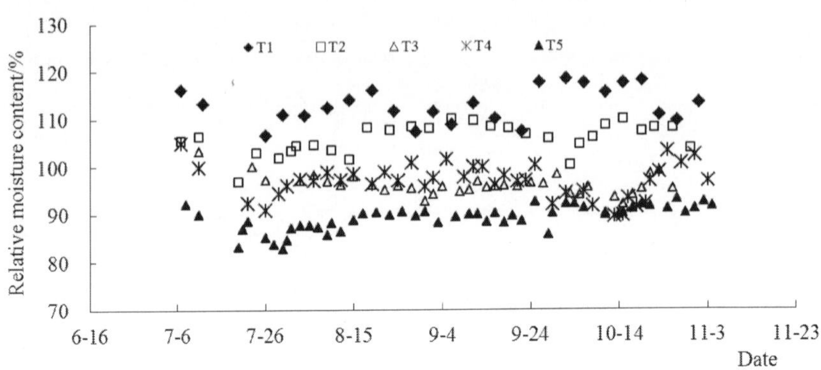

Figure 6. Moisture content after irrigation for all treatments

Refer to Table 7 for the substrate moisture contents analysis during the test. The average relative substrate moisture contents of T1, T2, T3, T4 and T5 are 98.5%θFC, 95.4%θFC, 90.4%θFC, 90.7%θFC and 84.7%θFC respectively. The ebb-and-flow is 5.7%~13.8%θFC higher than drip irrigation. Both of them have the phenomenon that the moisture content exceeds the substrate water holding rate. The moisture content of drip treatment T5 is lower than its substrate water holding rate. So 1/4 of the pot height is the critical value at which nutrient solution may affect the substrate moisture content. When the nutrient solution depth does not exceed 1/4 of the pot height, there is no large influence to the substrate moisture content, but when it exceed 1/4 of the pot height, the proportion of number of days when the moisture content exceeds 100%θFC increases along with increasing of the nutrient solution depth, while the number of days when it is between 80%~100%$\theta FC$is relatively decreasing.

The substrate moisture content is the main factor of influence for the substrate water, air and temperature environment. Because organic substrate is different from soil, absorption, research on storage and migration of water in substrate is not intensive yet. Qi Haiying (Qi Haiying et al., 2009) find that the substrate pore structure and volume are obviously affected by various water supply modes. Furthermore, pore structure transformation may cause change of water holding capacity. It may be drip arrow local water supply affecting the tested culture substrate volume and pore structure that cause the relative substrate moisture content is lower than the substrate water holding rate after drip irrigation (Ma Fusheng et al., 2012). It is necessary to study further for the law of the pore structure change for soilless culture substrate and its influence to the substrate water absorption, storage and dissipation under various water supply conditions.

Table 7. Level of substrate moisture content for all treatments

Moisture content	Item	Treatment				
		T1	T2	T3	T4	T5
>100%θ_{FC}	Number of days/d	54	43	7	6	0
	Percentage in total test days/%	43.5	34.6	5.6	4.7	0
80%~100%θ_{FC}	Number of days/d	70	81	117	118	124
	Percentage in total test days/%	56.5	65.4	94.4	95.3	100
Average substrate moisture content during the test/(%θ_{FC})		98.5	95.4	90.4	90.7	84.7

4.2.2. Irrigaiton timing and irrigation water use efficiency

In this paper, the irrigation periods of all treatments during the test are 5~7 d (Average 5.6 d) for T1, 4~6 d (Average 4.9 d) for T2, 3~5 d (Average 3.7 d) for T3 and T4, and 2~3 d (Average 2.6 d) for T5 respectively. Based on those, when all the low irrigation threshold are 80% of the substrate water holding rate, the irrigation period of the ebb-and-flow irrigation is longer than drip irrigation. The average irrigation periods of T1, T2, T3 and T4 are 3d, 2.3d and 1d longer than T5, the treatment of drip irrigation, which is in conformity with the law of substrate moisture content change. So 1/4 of the pot height is the critical value of the nutrient solution depth at which the ebb-and-flow irrigation period may be affected. When the nutrient solution depth does not exceed 1/4 of the pot height, there is no large difference among irrigation periods, but when it exceed 1/4 of the pot height, the irrigation period is prolonged along with increasing of the nutrient solution depth. Viz. under the conditions of controlling the same low irrigation threshold, ebb-and-flow irrigation may prolong the irrigation period and reduce manpower. Catherline A. Neal (Catherline A. Neal et al, 1993) find the irrigation period of ebb-and-flow irrigation is longer than drip irrigation, the same result as this paper.

As the main production form of facility flower, substrate potting mainly adopt walking tube-well irrigation, drip irrigation and other traditional water supply technology. The drip irrigation field water utilization coefficient of tested substrate in this paper is only 0.4~0.5 (Ma Fusheng et al., 2012), which is even lower than the limit of 0.9 specified in *Technical code for microirrigation engineering* (GB/T 50485-2009), there is serious clean and good water and fertilizer leakage loss. This may be affected by the dripper flow. According to Li Mingsi and others' (Li Mingsi et al., 2006) study, there is obvious influence to the soil wetting pattern. But the impact of law from the dripper flow on the organic culture substrate wetting pattern with finite volume needs to be study further. With the help of the ebb-and-flow irrigation technology, equipment such as residual water recycling, water treatment and recycling can be used to achieve circulation utilization of water and fertilizer resources, so the water and fertilizer utilization efficiency can be more than 90% (Zhang Xiaowen et al., 2011). To compare with drip irrigation, ebb-and-flow irrigation may improve water and fertilizer utilization efficiency, and reduce their leakage loss significantly.

4.2.3. Water consumption of anthurium

Refer to Fig. 7 for changing process of water consumption rate of anthurium in the 5 treatments. The water consumption rate of the tested anthurium increases gradually with fluctuation along with extension of growth and development. During the test, the Max. water consumption rates of T1, T2, T3, T4 and T4 are 52.3, 47.8, 46.8, 45.2 and 33.6 g/ (pot d) respectively, while the Min. water consumption rates are 12.3, 11.2, 11.3, 9.0 and 7.0 g pot^{-1} d^{-1} respectively, and the averages are 35.1, 31.7, 30.6, 29.8 and 21.1 g pot^{-1} d^{-1}. All the water consumption rates of ebb-and-flow irrigation treatment of anthurium are higher than drip irrigation.

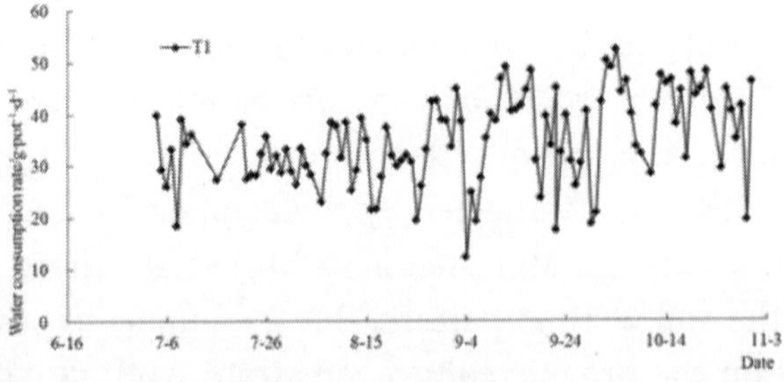

a. T1(Ebb-and-flow irrigation, the nutrient solution depth is 1/2 of the pot height)

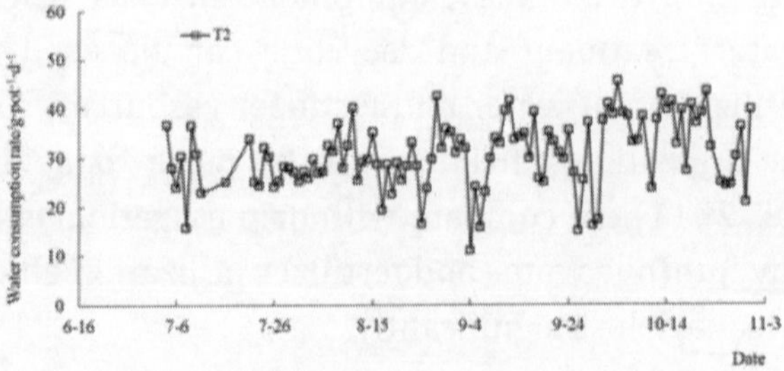

b. T2 (Ebb-and-flow irrigation, the nutrient solution depth is 1/3 of the pot height)

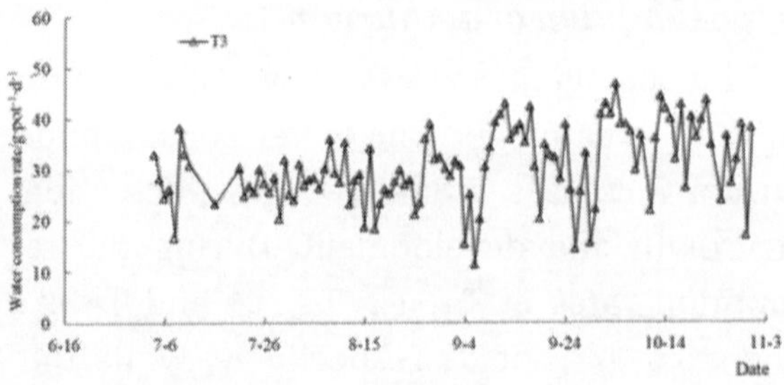

c. T3 (Ebb-and-flow irrigation, the nutrient solution depth is 1/4 of the pot height)

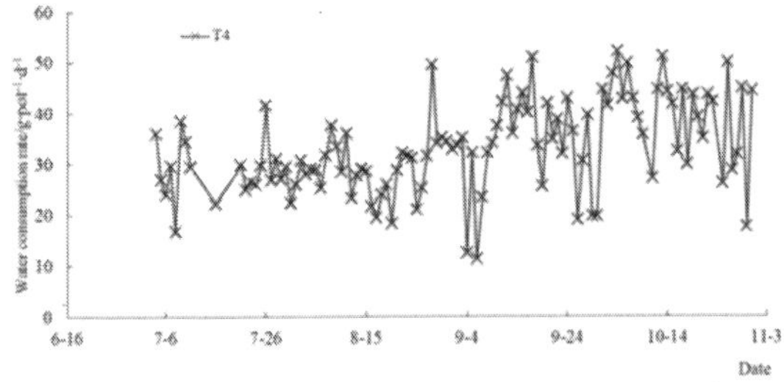

d. T4 (Ebb-and-flow irrigation, the nutrient solution depth is 1/5 of the pot height)

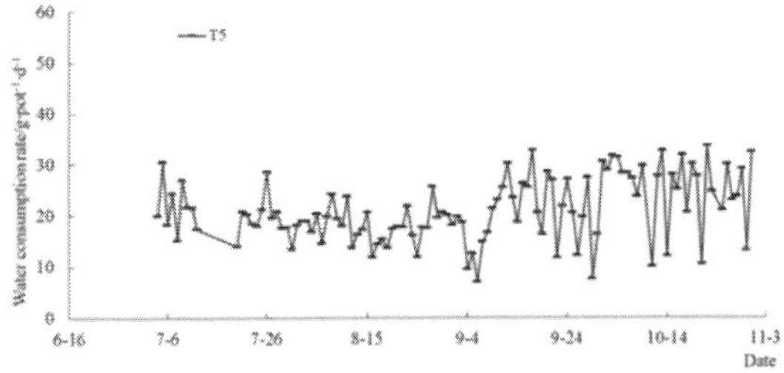

e. T5 (Drip irrigation)

Note: Count by pot.

Figure 7. Daily water consumption for all treatments during experiment

Refer to Table 8 for the monthly average water consumption rate and water consumption during the test. From July to October, the water consumption rate and water consumption of each treatment increase along with plant growth. The irrigation modes affect water consumption of anthurium obviously. The average water consumption rate, monthly water consumption and total water consumption of ebb-and-flow irrigation are higher than drip irrigation obviously during the test, the total water consumptions of T1, T2, T3 and T4 increase 69.3%, 53.2%, 47.5% and 44.2% than T5, so the variation trend of increasing along with increasing of the nutrient solution depth generally.If the nutrient solution utilization efficiency in ebb-and-flow irrigation is 0.9, and in drip irrigation is 0.5, nutrient solution used in ebb-and-flow irrigation during the test is 4 106.6~4 820 g pot^{-1} d^{-1}, 6%~20% lower than drip irrigation (5 124.6 g pot^{-1} d^{-1}), it is reduced distinctly.

Table 8. Daily and total water consumption amount of Anthurium for all treatments from 07-04 to 11-04

Item	Treatment	Time				
		07-04-07-31	08-01-08-31	09-01-09-30	10-01-11-04	Total
Average water consumption rate /($g \cdot pot^{-1} \cdot d^{-1}$)	T1	30.8	32.0	34.2	41.7	-
	T2	27.6	29.6	30.5	37.8	-
	T3	26.7	28.3	29.9	36.0	-
	T4	27.1	28.5	26.7	35.8	-
	T5	18.1	18.0	20.0	25.7	-
Water consumption/ ($g \cdot pot^{-1}$)	T1	861.8	991.4	1026.3	1458.2	4337.6
	T2	771.8	918.9	914.1	1321.8	3926.7
	T3	747.4	876.6	896.8	1259.5	3780.2
	T4	758.6	884.9	801.0	1251.4	3695.9
	T5	505.6	559.5	599.1	898.0	2562.3

4.2.4. Growth and quality of anthurium

The plant height and the crown diameter are the key indexes concerned ornamental quality of the anthurium, refer to Fig. 8 for the plant height and the crown diameter of the anthurium in each treatment. By the end of the test, the plant heights of T1, T2, T3, T4 and T5 are 33.6, 34.7, 35.3, 38.4 and 27.0 cm respectively, the crown diameters are 41.9, 42.6, 40.6, 43.9 and 33.3 cm respectively. The plant heights and the crown diameters from ebb-and-flow treatment are better than drip irrigation treatment obviously ($p=0.05$). The plant height of T4 in ebb-and-flow treatment is 14.3% higher than T1 ($p=0.05$), but the plant heights among other treatments and the crown diameters of all treatments have no obvious difference.

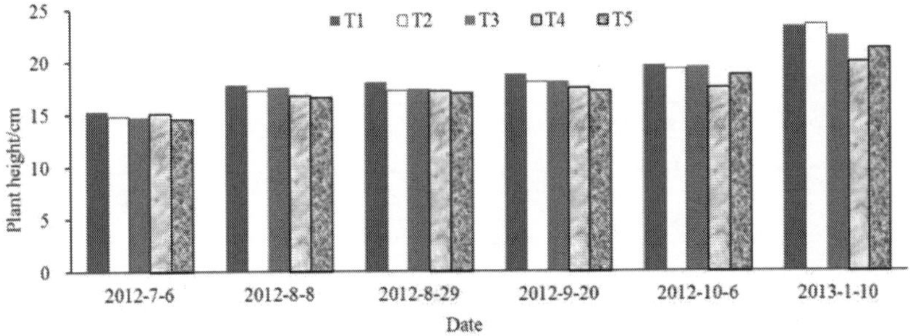

a Plant height

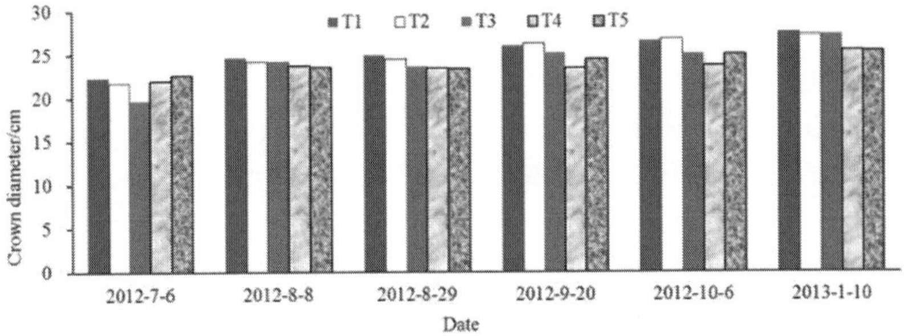

b Crown diameter

Figure 8. Plant height and crown diameter for all treatments

Refer to Table 9 for the spathe of each treatment. The spathe is the main ornamental part of the anthurium. By the end of the test, the quantity difference of the spathe in each treatment is not large, all are about 3. Compared with T5 of drip irrigation, the length of the spathe with ebb-and-flow irrigation increases 0.8~1.5 cm, the width of it increases 0.5~1.8 cm, except for the inflorescence height of T4 with ebb-and-flow irrigation treatment is about 1cm higher than T5, the inflorescence heights of other treatments have no large difference from T5. There is no obvious difference among treatments ($p=0.05$).

Table 9. Spathes of all treatments

Treatment	Parameters			
	Qty. of spathe/pc	Length of spathe/cm	Width of spathe/cm	Height of inflorescence/cm
T1	3.0	11.2	7.8	3.7
T2	2.6	11.5	8.1	3.6
T3	2.4	11.9	8.2	3.7
T4	3.4	12.9	9.1	4.4
T5	2.4	10.4	7.3	3.4

In conclusion, when the low irrigation threshold are controlled to 80% of the substrate water holding rate, growth of the anthurium is affected obviously from irrigation modes. Anthurium growth of ebb-and-flow treatment is better than that from drip irrigation treatment. The anthurium plant height and spathe size of T4 with the nutrient solution depth being 1/5 of the pot height appear uniform superiority.Catherine A. Neal (Catherine A. Neal et al. 1992) find plant growth is affected by irrigation modes, while ebb-and-flow irrigation is a technology that may improve irrigation water utilization efficiency and get potential optimal growth. Interaction between water and fertilizer is an important factor for plant growth. John M. Dole (John M. Dole et al, 1994) find the Poinsettias with ebb-and-flow irrigation has the best water utilization efficiency, plant height, stem diameter, leaf width and total amount of dry matter, in addition, its growth is affected by the nutrient concentration of the nutrient solution. Daniel I. Leskovar (Daniel I. Leskovar et al, 1998) find the irrigation mode, water regime and fertilizer application are key for growth of the plant root system and sprouting, but their interaction mechanism needs intensive study. Ma Fusheng (Ma Fusheng et al., 2011) find the lower irrigation limit affect anthurium growth even under drip irrigation conditions. Influence to anthurium growth from various irrigation modes in this paper may be caused by water and fertilizer coupling mechanism of each treatment, the influence mechanism from soilless culture water and fertilizer utilization to plant growth needs intensive study.

4.2.5. Suitable irrigation system for anthurium

According to experimental results, the quantity, length and width of the spathe, in addition to the inflorescence height of T4 with the nutrient solution depth as much as 1/5 of the pot height are the best. In addition, its water consumption is 15% less than T1. So it is the suitable nutrient solution control depth with ebb-and-flow irrigation, here the nutrient solution depth is 2.56cm, identify with the result of 25mm for most flower nutrient solution depth recommended by Yang Tieshun (Yang Tieshun et al., 2009). So in the testing conditions of this paper, in order to get optimal anthurium quality, the 1/5 pot height shall be taken as the ebb-and-flow irrigation nutrient solution depth, viz. 2.56cm, the irrigation period is affected by water, fertilizer, air and temperature environment of flower, it is 3~5d.

4.3. Conclusions

(1) The irrigation technologies are key factor to the substrate moisture content, the irrigation period and the water consumption law. The substrate moisture content after ebb-and-flow irrigation is between 96%~113%θFC, increase 7%~23%θFC than 89.1%θFC of drip irrigation; the average substrate moisture content during the ebb-and-flow test is 90.4%~98.5%θFC, increase 5.7%~13.8%θFC than 84.7%θFC of drip irrigation treatment. The anthurium water consumption with ebb-and-flow irrigation is 3 696~4 338 g pot^{-1} d^{-1}, increase 44.2%~69.3% than 2 562.3 g pot^{-1} d^{-1} of drip irrigation; the irrigation water utilization increases from 0.4~0.5 of drip irrigation to 0.9, the irrigation nutrient solution reduces from 5124.6g/pot to 4106.6~4820.0g/pot; the irrigation period of ebb-and-flow irrigation is 3~7 d, extending 1~3 d than drip irrigation in average, which improve the labor productivity.

(2) The irrigation modes have significant influence to quality of the anthurium. The plant height of the anthurium in ebb-and-flow irrigation treatment is 33.4~38.6 cm, 6.6~11.4 cm higher than 27.0 cm in drip irrigation, the crown diameter is 40.6~43.9 cm, 7.3~10.6

cm higher than 33.3 cm in drip irrigation; the length and width of the spathe are 11.2~12.9~cm and 7.8~9.1 cm, 0.8~1.5 cm and 0.5~1.8 cm higher than 10.4cm (Length) and 7.3cm (Width) in drip irrigation respectively, the inflorescence height of the spathe is 3.6~4.4 cm, 0.2~1.0 cm higher than 3.4cm in drip irrigation treatment.

(3) The 1/4 pot height is the critical value of the nutrient solution depth that affects the substrate moisture content, the anthurium water consumption and the irrigation period of ebb-and-flow irrigation. When the nutrient solution depth is not more than the 1/4 pot height, there are no large differences among the substrate moisture content, the anthurium water consumption and the irrigation period; but when it is more than the 1/4 pot height, the substrate moisture content, the anthurium water consumption and the irrigation period soar along with increasing of the nutrient solution depth.

(4) When the ebb-and-flow irrigation nutrient solution depth reaches the 1/5 pot height, the ornamental value of the anthurium is optimal, furthermore, the water consumption reduces 15% than that of T1 at which the nutrient solution depth is the 1/2 pot height. The total irrigation water consumption reduces 20% than drip irrigation. Here the nutrient solution depth is 2.56cm, the irrigation period is between 3~5d, the average value is 3.7d. So 2.56cm can be the better nutrient solution depth under the testing conditions.

5. EXPERIMENTAL RESEARCH ON SUITABLE DRIPPER DISCHARGE FOR POTTED ANTHURIUM WITH SOILLESS CULTURE

5.1. Experimental Plant and Design

The anthurium is chosen as the testing object; its variety is "Alabama", which belongs to *Araceae* and*Anthurium*. The culture substrate is PINDSTURP peat soil imported from Denmark, with the dry volume mass of 0.16g/cm^3, and the water holding rate (θFC) of

0.4315 (V/V). The tested anthurium is cultivated in plastic pot with 16cm of top diameter and 12.8cm of height, the volume is about 1.4L, and the substrate is filled to the pot evenly. Refer to Table 10 for testing treatment.

Table 10. Experimentle treatments

Testing stage	Dripper flow/(L/h)				
T1	T2	T3	T4	T5	
Whole growing period	0.55	0.95	1.1	2.2	3.8

Note: T1 adopts the plain end water separator configured 4 outlets with the flow of 2.2L/h, T2 adopts the plain end water separator configured 4 outlets with the flow of 3.8L/h, T3 adopts the plain end water separator configured 2 outlets with the flow of 2.2L/h, T4 adopts the 2.2L/h dripper, and T5 adopts the 3.8L/h dripper, the capillary tube after the distributive pipe connects the drip arrow.

In this paper, all treatment irrigation quota are controlled by the substrate water holding capacity as the upper irrigation limit, and 60% of the substrate water holding capacity as the lower irrigation limit. The test starts on June 6, 2012 and ends on January 12, 2013, takes 221d totally. The placing density of tested anthurium plantlet is 27 pot/m^2; and 23 pot/m^2 from August 9, 2012 to the end of the test. The plantlet is 16cm in height with uniform growth when transplanting. For all treatments, connect the drip arrow with uniform 1m long capillary tube, insert shallowly in the pot substrate from the side direction about 1cm within the center area of the root system. Place the pot on the bracket. Repeat 7 times per treatment. There are the protection lines at both sides of the monitored sample plant. The self-prime stabilized pressure pump with the pressure tank is used for driving; the hydraulicproportioning fertilization pump made by USA is used for fertilization with water.All test plots are distributed randomly, the treatment measures such as farming, fertilizing, and pest control are same.

5.2. Results and Discussion

5.2.1. Influence to water consumption of soilless potted anthurium from emitters with various dripper flows

The water consumption rate of tested anthurium was analyzed in this paper. Table 11 shows the water consumption rate of each treatment is between 8~24g pot^{-1} d^{-1} in June and July, and between9~32 g pot^{-1} d^{-1} in August and September, reaching the peak of water consumption 12~36g pot^{-1} d^{-1} in October, and falling back to 8~24g pot^{-1} d^{-1} in November and December. The change process from June to December is rise-fall. Refer to Table 11 for the average monthly water consumption rate and the water consumption per treatment.

Table 11. Daily and total water consumption amount of all treatments

Item	Treatment	Month							
		6	7	8	9	10	11	12	Total
Average water consumption rate/ $g \cdot pot^{-1} \cdot d^{-1}$	T1	16.6	17.3	20.5	20.0	22.6	17.3	17.1	--
	T2	16.0	16.9	19.4	19.0	21.9	16.8	17.3	--
	T3	16.0	17.0	19.7	18.9	22.7	17.0	17.3	--
	T4	15.9	15.7	17.7	17.4	21.5	16.4	16.7	--
	T5	15.8	16.1	18.9	17.7	21.0	16.6	16.7	--
Water consumption/ $g \cdot pot^{-1}$	T1	414	479	637	599	679	485	529	3823
	T2	400	466	609	571	657	471	536	3709
	T3	400	472	612	567	680	476	535	3743
	T4	397	437	548	522	643	459	517	3523
	T5	396	448	585	530	629	464	517	3569

According to Table 11, the water consumptions of each treatment in the whole growth period are 3823g pot^{-1} of T1, 3709g pot^{-1} of T2, 3743g pot^{-1} of T3, 3523g pot^{-1} of T4 and 3569g pot^{-1} of T5, the variation trend goes down along with increasing of the dripper flow. The water consumption of T1 with the Min. dripper flow is the highest, which is about 8% more than T5 (The Max. dripper flow) and T4 (The second largest dripper flow). This may be that the dripper with a small flow can establish a higher substrate moisture content, thereby creates more favorable transpiration environment. The water consumptions of T2 and T3 (Their dripper flows are not different greatly) are also close. Though the dripper flow of T4 is far lower than T5, there is no large difference between their water consumption. So increasing the anthurium water consumption by reducing the dripper flow has obvious effect only in a certain range and certain gradient. Wang Xiukang (Wang Xiukang et al., 2010) find the influence to corn root system from the dripper flow. When 1L h^{-1}~2.5L $h^{-1}$is chosen as the dripper flow, the dripper flow of 2L h^{-1}has obvious influence to spatial distribution of corn root system.

5.2.2. Influence To Irrigation Period Of Soilless Potted Anthurium From Emitters With Various Dripper Flows

Refer to Table 12 for influence of soilless potted anthurium irrigation period (*T*) from emitters with various dripper flows. The irrigation period goes up along with decreasing of the dripper flow, so the manpower cost can be saved.

Table 12. Effect of different dripper discharge to water use efficiency and irrigation cycle of Anthurium

Treatment	T1	T2	T3	T4	T5
T (d)	8.3	8.1	8.1	7.6	6.3

5.2.3. Influence to anthurium growth and quality from emitters with various dripper flows

There is no obvious difference on the tested anthurium spathe by the end of the test. Analyze the monitoring information of July 6, August 8, August 29, September 20, October 6 and December 31. Refer to Fig. 9 (a) for influence to tested anthurium plant height from various dripper flows, and to the crown diameter refer to Fig. 9 (b).

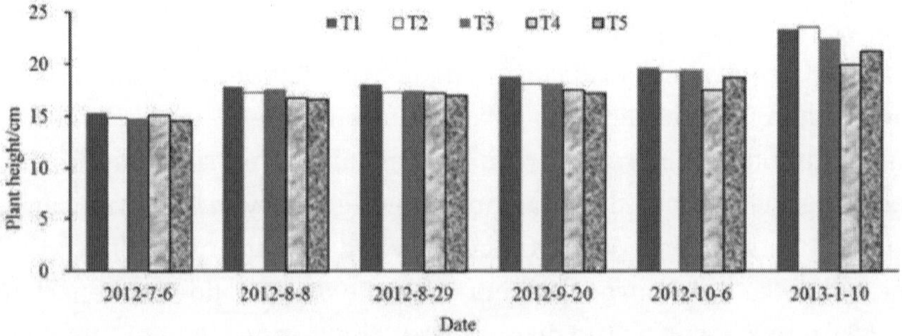

a Plant height

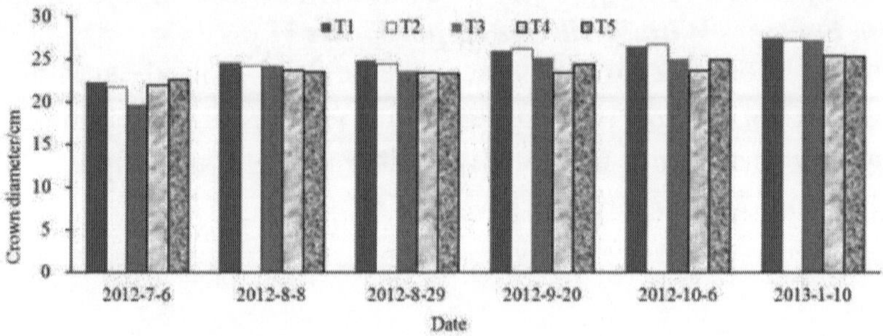

b Crown diameter

Figure 9. Effect of different low irrigation threshold to plant height and crown diameter

Based on Fig. 9, there is no large difference on the anthurium plant height and the crown diameter under the treatment conditions of T1, T2 and T3 (With lower dripper flows), all of them have obvious growth vigor. The plant heights and the crown diameters of T4 and T5 are lower than the first 3 treatments. By the end of the test, the plant heights of each treatment are 23.3cm of T1, 23.5cm of T2,

22.4cm of T3, 19.9cm of T4 and 21.2cm of T5; the crown diameters are 27.5cm of T1, 27.2cm of T2, 27.3cm of T3, 25.4cm of T4 and 25.3cm of T5.

5.2.4. Selection of drip irrigation emitters suitable for tested soilless potted anthurium

Based on comprehensive analysis on influence to the tested anthurium substrate moisture content, anthurium water consumption, field water utilization coefficient, irrigation period and anthurium quality from various dripper flows, to compare with T5, the anthurium water consumption of T1~T3 increases 5%~8%, and the irrigation period prolongs 1~2d, in addition to better ornamental effect on the anthurium plant height and crown diameter, especially the comprehensive advantage of T1. So within the range of the tested dripper flow, the dripper flow of 0.55L h^{-1} is recommended for tested soilless potted anthurium, with the irrigation period of 8.3d. Because of the limited research conditions in this paper, the suitability to tested anthurium irrigation if the dripper flow continues to reduce needs further investigation.

The low irrigation threshold in this paper is 60%θFC. Ma Fusheng (Ma Fusheng et al., 2012) and others find there is obvious law of influence to the field water utilization coefficient of soilless potted anthurium under the same dripper flow in various low irrigation threshold, the higher lower irrigation limit, the larger field water utilization coefficient. Selection and supporting technology of dripper flow under various low irrigation threshold needs further study. Good water and fertilizer resource leakage loss in drip irrigation is serious, and becomes major environmental pollution source, so it is urgent to research the water absorption and storage mechanism of soilless culture substrate, develop the substrate culture drip irrigation technical mode, develop the substrate materials with stable physicochemical properties, improve cyclic utilization of good water and fertilizer resource, improve the utilization efficiency of water and fertilizer resource meanwhile guarantee flower quality, in order to support the flower industry for the achievement of water-saving irrigation, good product quality, industrial environment protection and so on.

5.3. Conclusions

(1) The water consumption rate of the tested anthurium increases from 8~24g pot^{-1} d^{-1} in June to 12~36g pot^{-1} d^{-1} in October, and then fall back to 8~24g pot^{-1} d^{-1} in December; the water consumption of the whole growth period per treatment is 3569g pot^{-1}~3823g pot^{-1}, with the change trend of decreasing along with the increasing of the dripper flow.

(2) Under the 3 low-flow drippers with the flows of 0.55L h^{-1}, 0.95L h^{-1} and 1.1L h^{-1}, the plant height and the crown diameter of the anthurium have no large difference, but all of them are higher than the two treatments of 2.2L h^{-1} and 3.85L h^{-1}, which affects the spathe largely. So a dripper with the small flow is conductive to good quality.

(3) The small dripper flow may prolong the irrigation period for about 2d, and get better ornamental quality. To compare and select the tested drippers based on the economic benefit and water utilization efficiency, the small-flow dripper is suitable for the soilless substrate potted flower.

6. RESEARCH ON MOISTURE CHARACTERISTIC PARAMETERS FOR SOILLESS CULTURE SUBSTRATE

6.1. Test Design

Perform the single factor test with typical soilless culture substrates, and use sandy soil and sandy loam soil for the contrast test. There are 10 treatments, domestic peat (T1), vermiculite (T2), perlite (T3), peat imported from Germany (T4), domestic peat and vermiculite in

2:1 of mass ratio (T5), domestic peat and perlite in 2:1 of mass ratio (T6), domestic peat, vermiculite and perlite in 1:1:1 of mass ratio (T7), domestic peat, vermiculite and perlite in 3:1:1 of mass ratio (T8), sandy soil (CK1), sandy loam soil (CK2). Because perlite and peat imported from Germany have large particle, so the air-dried volumetric specific gravities when filling are 0.12 g/cm^3 and 0.15 g/cm^3 respectively, sandy soil and sandy loam soil are 1.4g/cm^3, the other treatments are 0.25 g/cm^3. Determine the moisture characteristic parameters such as the porosity, the permeability coefficient, the water characteristic curve, the diffusivity and so on for each treatment. Repeat 3 times per treatment.

6.2. Results and Analysis

6.2.1. Porosity

During practical production of seedling raised with substrate, the total porosity is normally 70%~90%, the water-air ratio is normally 2.0~4.0, which may meet demands of crop to moisture and oxygen (Li Douzheng, 2006). But a single substrate is hard to meet all those requirements at the same time. Refer to Table 13, the water holding porosity of vermiculite (T2) is large, viz. vermiculite has good water holding capacity, but poor aeration porosity, so its water-air ratio is unreasonable. The perlite (T3) has large aeration porosity, viz. perlite has good air permeability, but poor water holding capacity. So in the mixtures of T5, T6, T7 and T8, vermiculite is used to improve its water holding capacity, and perlite is used to improve the air permeability.

Table 13. Comparison of the different height containers

	T1	T2	T3	T4	T5	T6	T7	T8	CK1	CK2
Total porosity/%	80.11	81.10	82.41	81.82	78.09	82.45	78.42	84.73	50.74	40.55
Aeration porosity/%	18.38	14.91	37.10	18.01	19.51	21.90	20.70	21.90	2.84	1.94
Water holding porosity/%	61.73	66.19	45.31	63.81	58.58	60.55	57.72	62.83	47.90	38.61
Water-air ratio	3.35	4.44	1.22	3.54	3.00	2.77	2.79	2.87	16.87	19.90

Refer to table 13, the porosity of the substrate is higher than sandy soil (CK1) and sandy loam soil (CK2) obviously, this is due to the particle of sandy soil and sandy loam soil is small, and the unit weight is considerably higher than that of the substrate. Thereby the porosity of sandy soil and sandy loam soil is small. To compare with sandy soil and sandy loam soil, moisture in substrate is easier to be used by plant, to compare the above mentioned 4 mixed substrate, T8 has the Max. porosity of 84.7%, the water-air ratio of 2.9, so it has large total porosity and good air permeability, and a reasonable water-air ratio.

6.2.2. Analysis of permeability coefficient

The permeability coefficient is an important index to affect plant growth, a permeability coefficient suitable for plant growth depends on the soil type. Normally sandy loam soil, loam and clay are suitable for corn growth. The permeability coefficients of the 3 kinds of soil are 6×10^{-2}, 6×10^{-3} and 6×10^{-4}mm min^{-1}(Du Yanling et al., 1992). Currently, there is seldom research on the permeability coefficient of a substrate. If the permeability coefficient is too high, the substrate is difficult to hold water to cause leakage after irrigation and waste water resource, if the permeability coefficient too low, the air permeability of the substrate is bad, which may affect breath and growth of the root system (Shi Lianhui et al., 2008).

Refer to Fig. 10, except for T3 and CK1, the standard deviations of all treatments are all 0.2%~10% without large fluctuation. It means this determination method for the permeability coefficient can represent the permeability coefficient of a substrate. The permeability coefficient of sandy loam soil is 5×10^{-2} mm/min, which can meet requirement for corn growth. After comparing the permeability coefficients of CK2 and the substrate, the permeability coefficient of the substrate is higher than CK2 obviously for the aeration porosity of the substrate is larger, and water infiltration rate is faster than sandy loam soil after irrigation. So it may reach the corn water demand quickly, and little and frequent irrigation is appropriate. By comparing the permeability coefficients of substrates, the permeability coefficient of T2 is the largest, 2.20 mm min^{-1}, while T6 is the smallest. In addition, the permeability coefficient of a single substrate is higher than the composite substrate obviously, for there are more small particles in the composite substrate to blocking the pores in the substrate to slow down the permeation. So if a single substrate is used for culture and seedling, the irrigation frequency shall be adjusted reasonably.

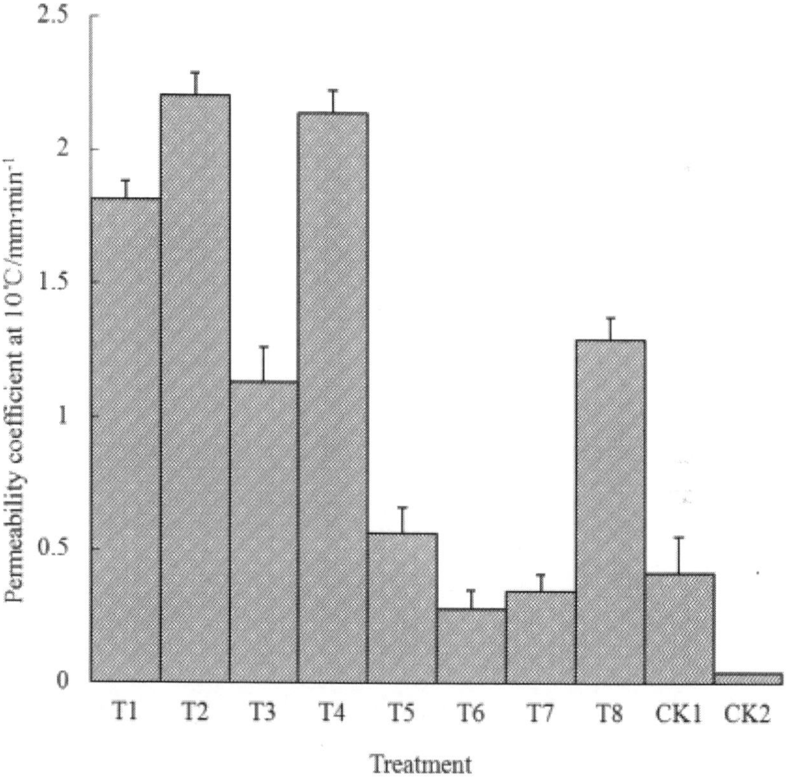

Figure 10. Comparison of permeability coefficient

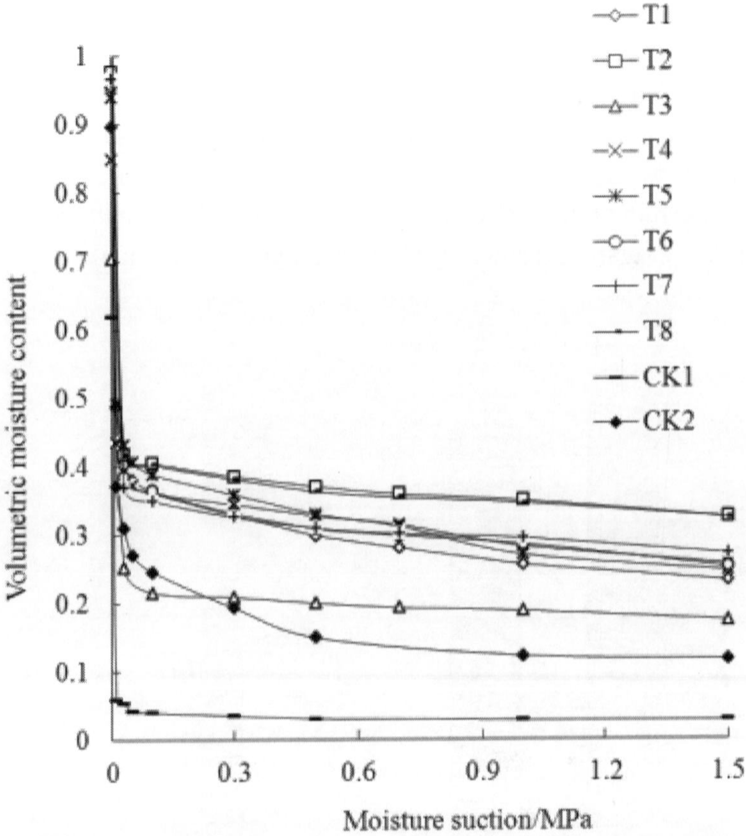

Figure 11. Comparison of moisture retention curves

In the drip irrigation engineering design, the dripper spacing is an important index to affect irrigation evenness, while the permeability coefficient is an important index to affect the dripper spacing (Yao Zhenxian et al., 2012). The dripper spacing suitable for the substrate shall be determined according to the dripper flow in order to improve the water utilization efficiency of the drip irrigation system.

6.2.3. Water characteristic curve

As shown in Fig. 11, at the lower suction stage at which the moisture suction is less than 0.1Mpa, the volumetric moisture contents of 10 treatments reduced sharply along with increasing of suction. The water holding capacity of T8 and T2 is the largest, and CK1 is the smallest. At the stage with the moisture suction is more than 0.1Mpa, the volumetric moisture contents reduce a little along with increasing of suction. Refer to Fig. 4 for the moisture content distribution, within the effective water range, viz. the substrate moisture suction is 0.01-0.1Mpa, the standard deviation of each treatment is 0.8%~3% with small fluctuation, it means the water characteristic curve can represent the water characteristic curve of the substrate.

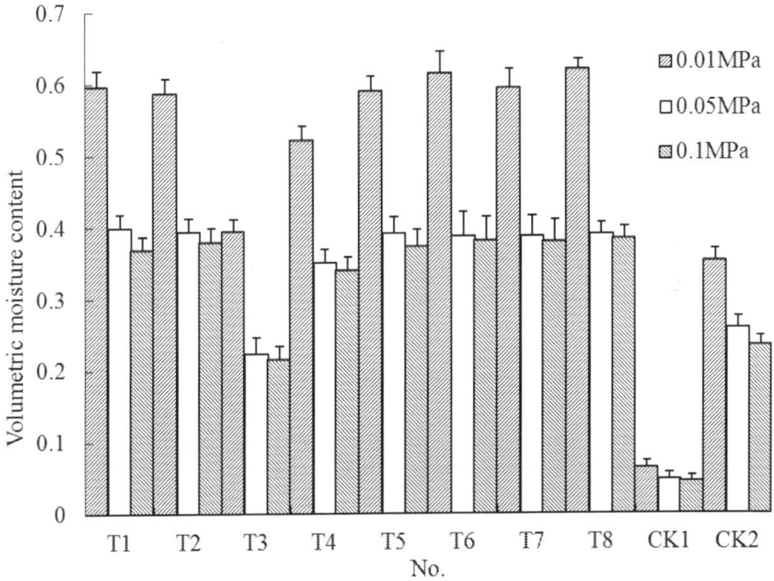

Figure 12. Comparison of water content between 0.01and 0.1 Mpa

In recent years, Kang Yuehu and others control the irrigation water suitable for growth and moisture efficient utilization of vegetables such as Chinese cabbage, cowpea and tomato under drip irrigation by controlling the soil matric potential. They guide irrigation of vegetableswithin the lower soil matric potential limit of -0.02~-0.05Mpa (Jiang Shufang et al., 2009; Zhang Chao et al., 2010; Wan Shuqin et al., 2009; Theodore W Sammis et al, 1980; Jia Junshu et al., 2011; Riviere L M et al, 2001). At the same time, take the volumetric moisture content within the ranges of 0.01~0.1Mpa, 0.01~0.05Mpa and 0.05~0.1Mpa of substrate water suction as the basis for division of substrate available water, easy available water and buffer water.

The available water, easy available water and buffer capacity are important indexes to judge water available for absorption by plant and guide the irrigation frequency (Shi Lianhui et al., 2008). The available water, easy available water and buffer capacity of 8 substrates may be obtained by the known water characteristic curve as Table 14 shown.

Table 14. Available water, easy available water and buffer capacity

	T1	T2	T3	T4	T5
Available water	22.7	20.8	17.9	18.1	21.7
Easy available water	19.6	19.3	17.1	17.1	19.8
Buffer water	3.1	1.5	0.8	1.0	1.9
	T6	T7	T8	CK1	CK2
Available water	23.3	21.3	23.5	2.0	11.7
Easy available water	22.6	20.6	22.9	1.7	9.3
Buffer water	0.7	0.7	0.6	0.3	2.4

As shown in Table 14, the available water capacity of substrate is higher than that of CK1 and CK2 obviously, which means the available water in the substrate is more higher, so crop may absorb more water in the substrate than in sandy soil or sandy loam soil. To compare the 8 substrates, both available water and easy available water in T8 are the highest, secondly is T6, which means the available water content in T8 and T6 is higher, so crop may absorb more available water. Their irrigation frequencies can be reduced than other substrates.

6.2.4. Analysis Of Horizontal Diffusivity

As Fig. 13 shown, when θ<0.3, the diffusivity changes slowly along with increasing of the moisture content, moisture in the substrate is mainly vapor movement [31] (Fan Yanwei, 2008); when θ>0.3, the diffusivity soars along with increasing of the moisture content, a higher substrate moisture content is in favor of substrate diffusive motion; when θ=0.3, refer to Table 3 for the diffusivity of each treatment.

When θ=0.3, the diffusivity of sandy soil is the largest, 654.75 $cm^2\ min^{-1}$, which means the horizontal suction and seepage velocity is quick, moisture spreads horizontally quickly; the second is perlite, 249.15 $cm^2\ min^{-1}$. Because perlite is lighter with smaller unit weight and larger particle size, moisture spreads quickly in perlite; to compare with sandy loam soil, the diffusivity of the substrate is higher, which means the horizontal diffusion velocity in substrate is quick. Thereby when using substrate as breeding and culture substrate, drip irrigation may improve the evenness and be in favor of crop water absorption.

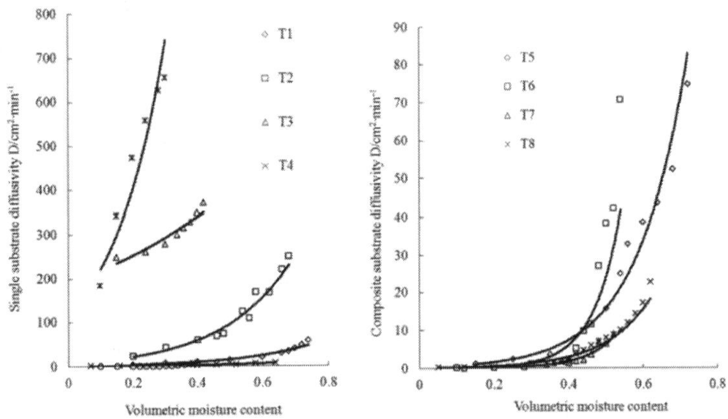

Figure 13. Correlation curve between diffusivity and water content

As Table 15 shown, fit the relation curve of diffusivity D and moisture content θ to get the formula of diffusivity D and moisture content θ.

Table 15. Comparison of diffusivity fitted formulas

No.	Fitting relation	Determination coefficient R^2
T1	$D(\theta)=1.6013e^{4.5724\theta}$	0.9771
T2	$D(\theta)=8.5301e^{4.8491\theta}$	0.9732
T3	$D(\theta)=188.64e^{1.4685\theta}$	0.9051
T4	$D(\theta)=0.6172e^{3.8262\theta}$	0.9804
T5	$D(\theta)=0.3662e^{7.5308\theta}$	0.9844
T6	$D(\theta)=0.0256e^{13.694\theta}$	0.9050
T7	$D(\theta)=0.013e^{12.3835\theta}$	0.9823
T8	$D(\theta)=0.1142e^{8.186\theta}$	0.9511
CK1	$D(\theta)=122.04e^{6.0059\theta}$	0.9155
CK2	$D(\theta)=0.0034e^{18.611\theta}$	0.9572

As shown in Table 15, the relationship between the horizontal diffusivity of substrate and moisture content of substrate are all match with the empirical formula

D(θ)=aebθD(θ)=aebθ

, changed as exponential function. The two have highly significant positive correlation relationship.

6.2.5. Comprehensive Analysis

Comprehensive assessment is performed to the 8 tested substrates with the matrix method, the standard is the larger total porosity (70%~90%), the higher score; the water-air ratio close to 3.0 is the best; because the substrate is hard to hold water, easy to leak after irrigation, the lower permeability coefficient, the higher score; for water availability, the larger substrate available water content, the higher score; since the diffusivity may affect water absorption by crop, the larger diffusivity, the higher score. Finally, comprehensive assessment is performed according to the synthesis score to various substrates; refer to Table 16.

Table 16. Matrix table

Influence factor	T1	T2	T3	T4	T5	T6	T7	T8
Total porosity	3	4	7	5	1	6	2	8
Water-air ratio	4	2	1	3	8	5	6	7
permeability coefficient	3	1	5	2	6	8	7	4
Water availability	6	3	1	2	5	7	4	8
diffusivity	6	7	8	4	5	2	1	3
Synthesis score	22	17	22	16	25	28	20	30

Accordingly, the synthesis scores of T6 and T8 are higher than other substrates obviously, which means T6 and T8 are better than others. Viz. the total porosity of T6 and T9 are larger, with good air permeability, reasonable water-air ratio, higher available water content, and in favor of crop absorption. Though the permeability coefficient and diffusivity of T8 are smaller, the irrigation evenness can be controlled by adjusting the irrigation frequency and irrigation amount.

6.3. Conclusions

1. By comparing the porosities of 8 substrates, the substrate formed by Domestic peat, vermiculite and perlite in a mass ratio of 3:1:1 has the largest total porosity, 84.7%; its water-air ratio is 2.9 with good air permeability and reasonable water-air ratio.

2. By comparing the permeability coefficients, the permeability coefficient of single substrate is higher than the composite

one. When a single substrate is selected for as the culture and breeding substrate, please pay attention to adjust the irrigation frequency reasonably.

3. By comparing the 8 substrates, the available water and easy available water capacity of the substrate with a mass ratio of 3:1:1 among domestic peat, vermiculite and perlite, and the one with a mass ratio of 2:1 between domestic peat and perlite are the highest, viz. the absorbable water for crop is high.

4. The relationship between the horizontal diffusivity of substrate and the moisture content of substrate are all match with the empirical formula $D(\theta)=ae^{b\theta}$, changed as exponential function. The two have highly significant positive correlation relationship.

5. According to the comprehensive assessment with the matrix method, the substrate mixed with domestic peat and perlite in a mass ratio of 2:1, and the substrate mixed with domestic peat, vermiculite and perlite in a mass ratio of 3:1:1 are the best. The total porosity of the two substrates are larger, with good air permeability, reasonable water-air ratio, higher available water content, and in favor of crop absorption, so they may be promoted and applied in production.

REFERENCES

1. Key soft subject research project of Beijing Municipal Science & Technology Commission, "Water-saving strategy research in suburb of Beijing", (Subject No.: Z0005170040231), Technical report [R]. Beijing Institute of Water Conservancy Scientific Research, 2007.

2. Guo Shenghu, Kong Dejie, Zhang Yuanpei, et al.,"Effects of different irrigation amounts on water consumption and water utilization efficiency of greenhouse

cucumber[J]". *Journal of Anhui Agricultural Sciences*, 2010, 36(34): 20568-20570. (In Chinese with English abstract)

3. Shen Xiaojun, Sun Jingsheng, Zhang Jiyang et al., "Research on water consumption law of winter wheat under deficit irrigation [J]". *Yellow River*, 2007, 29(11): 68-70.

4. Zheng Guobao, Kong Dejie, Zhang Yuanpei, et al. Effects of drip irrigation under film on water consumption of pepper in solar greenhouse[J]. Journal of Agricultural Sciences, 2010, 31(4): 50-52. (in Chinese with English abstract)

5. Zhao Ying, Li Bo, Wang Tieliang, et al. Study on water consumption of greenhouse eggplant under small tube flow irrigation[J]. Water Saving Irrigation, 2011(3): 33-35. (in Chinese with English abstract)

6. Tolga Erdem, Levent Arm, Yesim Erdem, et al. Yield and quality response of drip irrigated broccoli(*Brassica oleracea L. var. Italica*) under different irrigation regimes, nitrogen applications and cultivation periods[J]. Agricultural Water Management, 2010(97): 681-688.

7. Zhao Jinghua, Hong Ming, Ma Yingjie, et al. Journal of Irrigation and Drainage, 2010, 29(5): 94-97. (in Chinese with English abstract)

8. Liu Hongguang, He Xinlin, Wang Yaqin. Effects of regulated deficit irrigation on water consumption and yield of grape[J]. Journal of Irrigation and Drainage, 2010, 29(6): 109-111. (in Chinese with English abstract)

9. Abrisqueta J M, Mounzer O I, Ruiz-Sánchez M C. Root dynamics of peach trees submitted to partial root-zone drying and continuous def-icit irrigation[J]. Agricultural Water Management, 2008, 95(8): 959-967.

10. Gu Zhonghou, Tian Youliang, Guo Liansheng. Study on the sap flow dynamics and transpiration regulation of pinus tabulaeformis in Daqing mountains[J]. Forest Resources

Management, 2006(6): 57-59. (in Chinese with English abstract)

11. Philips N, Oren R, Zimmermann R. Radial patterns of xylem sap flow in none diffuse and ring porous tree species[J]. Plant Cell Environ, 1996(19): 983-990.

12. Sun Huizhen, Zhou Xiaofeng, Kang Shaozhong. Research advance in application of heat technique in studying stem sap flow[J]. Chinese Journal of Applied Ecology, 2004, 15(6): 1074-1078. (in Chinese with English abstract)

13. Ma Fusheng, Liu Honglu, Wu Wenyong, et al.Water consumption regularity of soilless culture Anthurium irrigated by drip irrigation in greenhouse[J].Journal of irrigation and darainge, 2011,30(6):61~64.

14. Jiang Desheng. *Modern horticultural media* [M]. Beijing: China Forestry Publishing House, 2006: 173-186.

15. Karam F, Saliba R, Skaf S, et al. Yield and water use of eggplants(*Solanum melongena L.*) under full and deficit irrigation regimes[J]. Agricultural Water Management, 2011(98): 1307-1316.

16. Guo Yuanyu as the chief editor. *Irrigation and drainage engineering* (Edition 3) [M]. Beijing: China WaterPower Press, 2007.

17. GB/T 50485-2009, *Technical code for microirrigation engineering*[S].

18. Qi Haiying, Zhang Shuxia, Song Chaoyu, et al. Effects of different substrates and watering modes on physicochemical properties of substrate and ornamental quality of Euphorbia pulcherrima[J]. Journal of Plant Resources and Environment, 2009, 18(3): 74-80. (in Chinese with English abstract)

19. Yang Wenbin, Hao Zhongyong, Wang Fengxin, et al. Effect of different irrigation thresholds on growth and yield of garland

chrysanthemum (*Chrysanthemum coronarium L.*) in greenhouse[J]. Transactions of the Chinese Society of Agricultural Engineering (Transactions of the CSAE), 2011, 27(1): 94-98. (in Chinese with English abstract)

20. Tian Yi, Zhang Yulong, Yu Na, et al. Effect of different low irrigation limit on growth,quality and yield of tomato under subsurface drip irrigation in greenhouse[J]. Agricultural Research in the Arid Areas, 2006, 24(5): 88-92. (in Chinese with English abstract)

21. Sun Huayin, Kang Shaozhong, Hu Xiaotao, et al. Response of greenhouse sweet pepper under alternate partial root-zone irrigation to different irrigation low limits[J]. Transactions of the Chinese Society of Agricultural Engineering (Transactions of the CSAE), 2008, 24(6): 78-84. (in Chinese with English abstract)

22. Li Xia, Xie Yingge, Xue Xuzhang, et al. Difference of transpiration, fresh matter accumulation and fruit yield for potted tomato under moisture content of medium[J]. Acta Horticulturae Sinica, 2010, 37(5): 805-810. (in Chinese with English abstract)

23. Steady and health development of flower industry in China – 2009 analysis of statistical data for flower in China. Refer to the website of State Forestry Administration. http://slzy.forestry.gov.cn/portal/main/s/2437/content-438856.html

24. Poole R T, Conover C A. Fertilizer levels and medium affect foliage plant growth in an ebb and flow irrigation system[J]. J. Environ. Hort, 1992, 10(2): 81-86.

25. Erin James, Mac van Iersel. Ebb and flow procuction of Petunias and Begonias as affected by fertilizers with different phosphorus content[J]. HortScience, 2001, 36(2): 282-285.

26. Catherine A. Neal, Richard W. Henley. Water use use and runoff comparisons of greenhouse irrigation systems[J]. Proc. Fla. State Hort. Soc, 1992(105): 191-194.

27. Chase R, Conover C A. Algae control in an ebb and flow irrigation system[J]. Proc. Fla. Stete Hort. Soc., 1993(106): 280-282.

28. Robert W. Rigsby. Ebb and flood irrigation system[P]. United States Patent: 5307589, May 3, 1994.

29. Ren Jianhua. "Water and nutrient solution ebb-and-flow irrigation"[J]. *Water-saving Irrigation*, 2004(3): 49-50.

30. Yang Tieshun. "Discussion on surface ebb-and-flow irrigation"[J]. *Agriculture Engineering Technology* (Greenhouse Horticulture) , 2009(4): 23-24.

31. Zhao Chanli, Gao Yuting. "New ebb-and-flow irrigation and cultivation technology settled in Ningxia"[N], *Huaxing Times*. August 19, 2010. Edition 002.

32. Zhang Xiaowen, Tian Zhen, Liu Wenxi, et al. Research and extension of Ebb-and-flow irrigation system[J]. Transactions of the Chinese Society of Agricultural Engineering (Transactions of the CSAE), 2011, 1(1): 80-83. (in Chinese with English abstract)

33. Zhang Li, Wang Yong. Primary research of tidal irrigation potted *Hydrangea macrophylla*(Thunb.) seringe[J]. Northern Horticulture, 2011(20): 77-79. (in Chinesewith English abstract)

34. GB/T 50485-2009, *Technical code for microirrigation engineering*[S].

35. GB/T 50485-2009, Technical code for microirrigation engineering[S].

36. Li Mingsi, Kang Shaozhong, Sun Haiyan. Relationships between dripper discharge and soil wetting pattern for dripirrigation[J]. Transactions of the Chinese Society of Agricultural Engineering (Transactions of the CSAE), 2006, 22(4): 32-35. (in Chinese with English abstract)

37. John M. Dole, Janet C Cole, Sharon L von Broembsen. Growth of Poinsettias, nutrient leaching, and water-use efficiency respond toirrigation methods[J]. Hortscience, 1994, 29(8): 858-864.

38. Daniel I. Leskovar. Root and shoot modification by irrigation[J]. HortTechnology, 1998, 8(4): 510-514.

39. Meng Qingguo, Chen Fengming, Liu Yingjie. "Application of drip arrow microirrigation technology on pottedflower production" [J]. *China Flower & Horticulture*, 2005, (14): 50-52.

40. Pang Yun, Zhang Junsheng, Zhang Zhi, Zhang Wenzhu. "Comparative test between various water-fertilizer treatment and soil culture border irrigation for soilless culture tomato under drip irrigation"[J]. *Acta Agriculturae Boreali-sinica*, 2006,21 (Special) : 41-43.

41. Yue Bin, Wang Liuyun. "Test research on drip irrigation soilless culture tomato"[J]. *Water-saving Irrigation*, 1998(2): 18-1948.

42. Wang Ronglian, Yu Jian, Tan Yumei et al. "Influence to yield of soilless culture cherry and tomato from water and fertilizer coupling during drip irrigation in greenhouse"[J]. *Journal of Irrigation and Drainage*, 2009, 28(4): 87-89.

43. M.K. Merica, I.H. Tuzelb, Y. Tuzelc, G.B. Oztekinc. Effects of nutrition systems and irrigation programs on tomato in soilless culture [J]. Agricultural Water Management, 2011(99): 19-25.

44. D. Massaa, L. Incroccib, R. Magginib, et al.Strategies to decrease water drainage and nitrate emission fromsoilless cultures of greenhouse tomato[J]. Agricultural Water Management, 2010(97): 971–980.

45. Zhong Gangqiong, Chen Yongbo, Long Jun et al., "Research on irrigation amount and irrigation mode for konjac soilless culture"[J]. *Agricultural Engineering Science*, 2005,21(3):340,341&353.

46. Wang Xiukang, Li Yuannong, Xing Yingying et al. "Influence from dripper flow to summer maize root system"[J].*China Rural Water and Hydropower*, 2010 (11) :66-68.

47. Tian Jilin, Wang Yinhu. Current situations and prospects of researches on soilless-culture substrates[J]. Acta Agriculturae Shanghai,2000,16 (4):87~92

48. Li Shijun, Gao Lihong, Zhuang Zhonglian. Achtievement new technology of the recent past in soilless culture in China and its trend[J].Journal of Changjiang Vegetables,1997,(5):1~4.

49. De Boodt M, Verdonck O. The Physical Properties of the Substrates in Hort[J] . Acta Hort, 1983, 26: 37~44.

50. Zhou Yanli, Cheng Zhihui, Meng Huanwen, et al. Effects of organic substrate compositions on growth yield and quality of tomato[J].Jour. of Northwest Sci-Tech Univ. of Agri. And For.:Nat.Sci.Ed., 2005,33(1):79~82

51. Li Jingrui, Zhang Tiejun, Gao Hongbo,et al.Effects of different substrates on growth and chlorophyll fluorescence characters of eggplant seedlings[J].Acta Agriculturae Boreali-occidental is Sinica, 2010,19(6):139~143.

52. Wang Yajun, Wei Xinghu, Xie Zhongkui, et al.Effects of different culture medium on growth and bulb of cutting flower lily[J]. Acta Agriculturae Boreali-occidental is Sinica, 2003,12(4):109~112.

53. Zhang Jianhua, Deng Pu. Effects of different culture medium on survival rate and growing situation of anthurium and raeanum tissue culture seedling after transplanting[J]. Journal of Shanghai Jiaotong University:Agricultural Science, 2009,28(6):624~626(in china)

54. Zhang Ning, Feng Mei. Ping Jicheng. Effect of different substrates on growth of strawberry[J].Northern Horticulture, 2011,(16):51~53. (in chinese)

55. R. Kasten Dumroese, Juha Heiskanen, Karl Englund,et al. Pelleted biochar: Chemical and physical properties show potential use as a substrate in container nurseries[J]. Biomass and Bioenergy, 2011,(35):2018~2027.

56. David W R. Water, media and nutrition for greenhouse crops[M]. Batavia,Illinois USA: Ball Publishing, 1996.

57. Tang Jing, Kang Hongmei.Physical and chemical properties of several important media[J].Chinese Journal of Soil Science,2006,37(2):291-293(in chinese)

58. Zhao Jiuzhou, Chen Jiemin, Wang Yiqing, et al.Studies on the physical and chemical characters of six-soilless-media[J].Journal of Laiyang Agricultural College, 2001,18(3): 161~164,173. (in chinese)

59. Li Jing, Zhao Xiulan, Wei Shiqiang, et al.Study on the physico-chemical properties of soilless cultural substrates of pollution-free vegetable[J].Journal of Southwest Agricultural University, 2000, 22(2):112~115. (in chinese)

60. Chen Zhende, He Jinming, Huang Junjie, et al. Composition and Physical and Chemical Characteristics of Plug Seedling Media of Vegetables[J].Transaction of the CSAE, 1998,14(2):192~197. (in chinese)

61. Tian Jilin, Xi Zhenbang, Chen Chunhong. Study on the feasible parameter of substrate quality for soilless culture[J].Acta Agriculturae Shanghai, 2003,19(1):46~49. (in chinese)

62. Chen Lixin. *Coursebook for Soil Experiment and Practice*[M].Harbin: Northeast Forestry University Press, 2005: 59~63.

63. Duan Zhaolin. Application on the matrix of environmental impact comprehensive assessment based on the principle of getting value from environmental elements proportion value[J].Yunnan Environment Science, 2003,22(4):59~62. (in chinese)

64. Li Douzheng. *Study on effects of components and particle sizes on pore properties of media*[D]. Shandong Agricultural Unverisity, 2006. (in chinese)

65. Du Yanling, Xu Guo'an. *Finite element method and electric network method for seepage analysis*[M].Beijing: Water Conservancy and Electric Power Press, 1992.

66. Shi Lianhui, Zhang Zhiguo Liu Dengguo et al. Comparison of physiochemical properties between spent mushroom compost and peat substrate and adjustment[J]. Transaction of the CSAE,2008,24(4):199~203. (in chinese)

67. Yao Zhenxian, Zhang Dapeng, Yao Miaosen. Breslers's mathematical model for determination distance between emitters by soil hydraulic parameter[J]. Transaction of the CSAE, 2012, 2(2):65~68. (in chinese)

68. Jiang Shufang, Wan Shuqin, Kang Yuehu.Effects of soil matric potential on Chinese cabbage yield and water use efficiency under drip irrigation[J].Water Saving Irrigation, 2009,(10):20~22. (in chinese)

69. Zhang Chao, Kang Yuehu. Wang Shuqin, et al.Effects of soil matric potential on cowpea yield and irrigation water use efficiency under drip irrigation[J].Journal of Irrigation and Drainage,2010,29(4):30~33. (in chinese)

70. Wang Shuqin, Kang Yuehu, Liu Shiping.Effects of soil matric potential and topping on tomato growth and water use

efficiency under drip irrigation[J]. Journal of Irrigation and Drainage, 2009,28(2):1~4. (in chinese)

71. Jiang Shufang, Wang Shuqin, Kang Yuehu, et al. Effects of soil potential on lettuce growth and water use efficiency under drip irrigation in semi-humid regions of north China[J].Journal of Water Resources and Water Engineering, 2011,22(6):72~76.

72. Theodore W Sammis. Comparison of sprinkler, trickle, subsurface, and furrow irrigation methods forrow crops[J]. Agron J, 1980,72:701~704.

73. Jia Junzhu, Kang Yuehu, Wang Shuqin, et al. Effect of soil matric potential on lycium barbarbum L. growth under drip-irrigation[J]. Journal of Irrigation and Drainage, 2011,30(6):81~84.

74. Rivière L M, Caron J. Research on substrates: state of the art and need for the coming 10 years[J]. Acta Hort, 2001,548:29~42.

75. Fan Yanwei.Study on numerical simulation of interference infiltration characteristic in film hole irrigation[D]. Yangling: Northwest Agriculture and ForestUniversity, 2008.

2

SUSTAINABLE WATER MANAGEMENT FOR URBAN AGRICULTURE, GARDENS AND PUBLIC OPEN SPACE IRRIGATION: A CASE STUDY IN PERTH

Raju Sharma Dhakal[1*], Geoff Syme[1], Edward Andre[1], Charles Sabato[2]

[1]Centre for Planning, Edith Cowan University, Perth, Australia

[2]Water Corporation, Perth, Australia

ABSTRACT

Urban agriculture has been increasingly popular as a form of modern agriculture in urban settings. It includes community gardens, fruit orchards, home gardens, veggie patches, public open spaces, reserves, urban forest, and recreational landscaping. However, irrigation using urban water supply has been identified as a major constraints for the development of urban agriculture. This study presents a sustainable water management trial at Butler, a northern sub-urban development in Perth, Western Australia, for urban irrigation. The trial system consists of a number of water saving features including untreated fit-for-purpose groundwater supplied via a third pipe network, drip irrigation, local weather station, soil moisture sensors connected with a local weather station, night time irrigation, soil enhancement with conditioning and mulching, and use of native plants and vegetation. The trial outcome was compared against controlled areas in terms of irrigation efficiency and sustainable water management for urban agriculture. The study demonstrated that a fit-for-purpose irrigation along with water sensitive land management could be a sustainable alternative for urban agriculture that would achieve a significant water saving and irrigation efficiency at urban settings. However, quality of untreated groundwater can be an issue while utilizing it for irrigation, but the research has shown that it can be managed with innovative irrigation techniques. This indicates that the fit-for-purpose irrigation system with water sensitive land management practices would be highly supportive in sustainable development of urban agriculture, vegetation and recreational landscaping.

Keywords: Urban Agriculture, Sustainable Irrigation System, Integrated Land and Water Management

1. INTRODUCTION

Urban agriculture has become increasingly popular across the world that includes a variety of activities: community gardens and fruit orchard, home gardens and veggie patches, urban forest, public open spaces, reserves, urban forest and recreational landscaping [1] -[5] . Urban agriculture differs from traditional agriculture as it is integrated into densely populated areas with limited land for food production and recreation space [4] . Urban agriculture can bring diverse vegetative structures back into urban system, support local bio-diversity, and provide ecosystem services across fragmented habitats and spatial levels [6] that in turn can reduce the impact of climatic variability [7] . In addition to food production, urban agriculture can cater for a wide range of urban community needs, including cultivation of vegetables, medicinal plants, spices, mushroom, fruit trees, ornamental plants, and other productive plants [5] [8] . The production of crop and agricultural goods, within and around cities, with a motivation of personal consumption or income generation [9] [10] thus integrates the local, urban and suburban economic and ecological system [1] .

Urban agriculture has a vital role in enhancing food production and bio-diversity, but that comes with significant costs and constraints [11] [12] . Albeit not required, urban agriculture utilizes significant amount of scheme water for growing foods and crops, which is usually expensive than agricultural water supplies [4] [7] . On the other hand, reducing outdoor use of scheme water for agriculture activities is a major component of many strategies to reduce urban water use and ensure reliable indoor water supply [4] . For instance, almost 40% of the scheme water is consumed for agricultural irrigation activities in urban and regional councils of Perth metropolitan. In drought period, the agricultural irrigation is normally restricted with a variety of bans (daytime and winter sprinkler ban), and demand management practices [13] -[16] . As a result, urban agricultural activities are not always supplied with sufficient irrigation, especially when they require most, hence affecting the continuity of food production and other agricultural activities. This condition is expected to be continued in WA, since a gap between water demand and supply is predicted to be 120 GL by 2030 [15] . In this context, an alternative irrigation system utilizing locally available fit-for-purpose water sources (e.g. groundwater) could have multiple benefits from helping water authorities in augmenting water supply and ensuring cheap and reliable irrigation for urban agricultural activities.

Previous studies in Australia and other countries have shown that alternative irrigation systems using groundwater and recycled wastewater are widely accepted for urban and regional agriculture, forestry and recreational landscaping [7] [17] -[22] . Wong [19] suggests that agriculture, garden and park irrigation are important avenues to utilize recycled water and nutrients within the build environment. This supports the integrated land and water management practices at respective levels by reducing water import into and net discharge out of the environment. Further, well-established urban agriculture activities would have potential to mitigate and adapt the climate change by reducing the urban heat island effects, managing micro-climate, and enhancing biodiversity and ecology [7] .

Similar studies have shown a growing acceptance among Australian communities towards alternative irrigation systems [17] [18] [22] -[24] . Hurlimann [17] observed that the majority of Mawson Lake community in South Australia accepted the irrigation use of recycled water as a better option compared to drinking water for their home gardens, veggie patches and other urban agriculture activities. Similarly, Australia-wide case studies conducted by Davis and Farrelly [22] -[24] have shown that a number of alternative water systems can be best utilized for efficient irrigation system. In context of WA, a number of studies consider groundwater as a reliable and sustainable source for outdoor irrigation activities [16] [18] [25] -[27] . These studies indicate that using groundwater as an alternative source for urban agriculture is technically feasible and receiving growing community acceptance all over Australia.

In this context, a suburban area in Perth ("The Green") was selected for this study that has alternative irrigation system via third pipe network. The system utilized local groundwater from shallow aquifer (20 - 60 m below the ground surface). 5 communal bores abstract the groundwater and supply to the irrigation ring without any treatment for irrigating the private gardens and the public open space in every alternate day [28] . This innovative irrigation system was investigated in terms of irrigation efficiency and its utility for urban agriculture. For that, total household drinking water and irrigation consumption, and total public open spaces usage in "The Green" were analyzed and compared with that of the surrounding suburban areas and Perth metropolitan. Further, public acceptance towards the groundwater as an alternative irrigation source was analyzed.

2. DESCRIPTION OF THE STUDIED SYSTEM

This study was carried out in "The Green" suburban community at Butler, Western Australia (Figure 1), where urban agricultural practices and recreational landscaping were being irrigated with an efficient irrigation system. Butler is a Greenfield suburban area of City of Wanneroo, which is 35 KM north of Perth CBD. The development of "The Green" began in 2005 and still to finish. "The Green" was compared as an experimental area with its neighbouring suburban development, mainly Ridgewood as control and Perth metropolitan city. Ridgewood is an established metropolitan suburban area having similar geography, climate, and socio-demography; but lacking the groundwater irrigation system and having a separate activity locus.

Figure 1. Location of the research area (Source: Contrywide Publication, Google Maps).

The irrigation system utilised the local groundwater without treatment, hence only used for irrigating home gardens, veggie patches, fruit orchards, community gardens, verges, public parks and ovals, urban forest, plantations, reserves, and bushlands [22] [29] .

The irrigation system incorporated water smart technologies and water sensitive land management practices for the efficient consumption and optimisation of the groundwater supplied for outdoor agricultural activities, which include:

1) Auto-operated irrigation system

a) Weather station;

b) Soil moisture sensors.

2) Irrigation techniques for achieving efficiency and reducing human contacts

a) Night time irrigation;

b) Drips and sprinklers irrigation.

3) Soil conditioning and mulching

4) Water efficient agricultural activities at home

5) Water efficient agriculture, forestry, and recreational landscaping at community

2.1. Automatic Irrigation System

All agricultural and outdoor activities in "The Green" were automatically supplied with the local groundwater on the basis of local weather condition. Real time weather information was obtained from a local meteorological station (Figure 2(a)); such as temperature, rainfall, relative humidity, wind velocity, etc. This information was coupled with soil moisture information received from in-situ soil moisture sensors (Figure 2(b)) across "The Green" in controlling and operating the irrigation system. Thus established automated irrigation system not only provided irrigation virtually always when required, but also optimized the irrigation supply by avoiding overwatering, and irrigation after rain.

2.2. Efficient Irrigation Techniques

The gardens, parks, and urban agricultural activities in "The Green" were provided night time irrigation (10 p.m.-6 a.m.) that avoided evaporation loss, and also reduced possible health hazards to humans a direct contact with the groundwater. A number of efficient irrigation techniques were utilised, such as: unique pop-up sprinklers, and subsurface drip reticulation. The sprinklers were designed for spraying bigger droplets that would avoid wind-drifting and supplying enough water even in shorter operation windows. Similarly, subsurface drip- pipes (Figure 3(a)) supplied irrigation near the root zone of crops that not only improved irrigation efficiency but also reduce evaporation loss.

2.3. Soil Conditioning and Mulching

When preparing the garden beds and lands for other agricultural activities, the soil condition was improved with sufficient soil mixture, soil conditioner, and mulch as shown in Figure 3(a) and Figure 3(b). This would not only enhance the soil quality, fertility, as well as the water holding capacity but also reduce the evaporation loss from the surface.

2.4. Water Efficient Agricultural Activities at Home

Water efficient food production and gardening activities were encouraged at each household to reduce irrigation use. As shown in Figures 4(a)-(c), vegetable gardens and front gardens were established with less water demanding crops, vegetable, hardy turf, and bush. Mostly native plants were used, lawn areas were reduced to improve irrigation efficiency.

(a) (b)

Figure 2. A local meteorological station (a) and soil moisture sensor (b) associated with the third-pipe irrigation system in "The Green".

(a) (b)

Figure 3. Photographs of garden bed preparation (a) and mulching (b) in "The Green".

2.5. Water Efficient Agriculture, Forestry, and Recreational Landscaping at Community

In "The Green", community gardens, public parks, and recreational landscaping were designed in water efficient way. Usual sprinklers were replaced with subsurface drip irrigation practices (Figure 3(a) and Figure 5(c)) and the lawn areas were dropped and replaced with less water demanding natives and bushes (Figure 5(a)). Trees and bushland (Figure 5(b)) were retained in the parks, so that these parks could be utilized for multiple purposes, such as: playground and recreational activities, community gardening, improving bio-diversity, urban cooling, and community appearance.

Furthermore, a higher density development was adopted in "The Green" that reduced residential garden areas and eventually the residential irrigation demand. The reduced residential gardens were, however, compensated with increased community gardens, public open spaces and reserves [28] [30] . In addition, wide range of water sensitive land management practices were adopted, namely: porous pavement, bio-retention trenches and basins, grassed swales, terraced gardens, and artificial water bodies [22] . These practices would help reducing irrigation demand, improving irrigation efficiency, facilitating aquifer recharge, and recycling urban water and nutrients within the build environment. These innovative designs facilitated integrated and sustainable water management at local and urban level. The innovative irrigation and water sensitive land management practices were expected to be an important avenue for ensuring efficient and integrated land and water management [14] , and improving neighbourhood appearance and liveability [31] at various spatial level.

Figure 4. Photographs of veggie patches (a) and (b), and home garden (c) in "The Green".

(a) (b) (c)

Figure 5. Public parks with (a) native plants, (b) trees and bushland, and (c) recreational landscaping in "The Green".

3. RESEARCH METHOD

A case study with mixed methodology was adopted to evaluate the agricultural irrigation efficiency and water conservation as a result of the fit-for-purpose irrigation system and water efficient urban development. This paper utilizes the secondary data on drinking water consumption, irrigation usage in gardens and parks and land use pattern to analyze the scheme water saving and irrigation efficiency at "The Green". The drinking water consumption and irrigation usage data were obtained from the Western Australian Water Corporation and land use data were obtained from the City of Wanneroo. At the same time, the qualitative information from survey and stakeholders interviews were analyzed to understand community acceptance and socio-economic challenges associated with the alternative irrigation system. The data were sorted, managed and analyzed using computer software: Microsoft Excel 2007, IBM SPSS 22, IBM SPSS Amos 22 and NVivo 9.

The drinking water supply was metered and measured in KL/year/household, whereas the groundwater irrigation supply was not metered. Hence a bulk amount of residential groundwater consumption was divided by the number of households connected to the irrigation system to get the average groundwater usage (in KL) per household per year. Similarly, the annual public open space irrigation usage was measured with the meter installed in each public parks and verges and given in ML/year.

The total household water usage was calculated as a sum of average drinking water consumption and irrigation usage. The comparative analysis of drinking water consumption data across the studied areas explained the drinking water conservation, whereas the analysis of total domestic water (Drinking and groundwater) usage data provided the net water efficiency situation at "The Green". The analysis of drinking water conservation was not restricted only for the survey participants. It utilized the data for the whole studied suburban areas to explore a complete picture of water conservation and irrigation efficiency scenarios.

4. RESULTS AND DISCUSSIONS

4.1. Water Conservation Scenario

In "The Green", both drinking water and groundwater irrigation were supplied; whereas in Ridgewood and Perth metro region, only drinking water was supplied. "The Green" was started since 2007 but the number of households connected to the groundwater irrigation system was very low until 2010, hence only the data from 2011 were used for the analysis. In Table 1, the total household water consumption in "The Green" (drinking water and groundwater irrigation) is compared with that of Ridgewood and Perth metropolitan (only drinking water).

The average drinking water consumption in "The Green" was significantly lower compared to Ridgewood and Perth metropolitan for all times. A typical household in "The Green" consumed only half of the drinking water used by an average Perth Metropolitan household. Such a significant water savings in "The Green" was mainly because of groundwater irrigation system that replaced the drinking water usage in garden irrigation and other outdoor agricultural activities.

As shown in Table 1, the total household water consumption in "The Green" in 2012 appeared to be significantly higher than neighbouring suburban areas and metropolitan average. It could be attributed to increased usage of groundwater (188 KL) for new garden establishment, and other once-off usage such as construction works. It became further evident from continuous reduction in total water consumption in next two consecutive years (Table 1).

It should be noted that a newly established garden or lawn in Perth metropolitan could get irrigation exemption for up to 42 days in summer and 35 days in winter. During exemption period, a significant amount of water (10 mm/day in summer, and 7.5 mm/day in winter) would be consumed [34] . This was supposed to be the main reason for the higher amount of groundwater consumption in "The Green" given the number of newly established garden and lawns. This figure could get higher, but the water efficient irrigation system and land management practices help to restrain it.

As shown in Figure 6, the household irrigation consumption in "The Green" appeared to be significantly higher than allocated, whereas the public parks irrigation consumption figures were far below than the allocation limit. This could be linked again to the large number of irrigation exemptions for newly established gardens across "The Green" suburb in 2013, which was supported with the reduced consumption in successive years (Table 1). Whereas the groundwater consumption for public open space irrigation in "The Green" was significantly lower than the allocation limit as well as the average public open space irrigation in surrounding suburban areas and metropolitan figures. This indicates that the water efficient irrigation technology and the water sensitive land management practices could significantly reduce the irrigation demand of public parks, verges, and other green infrastructures at the urban setting. The findings supports the empirical ideas on irrigation efficiency and urban water conservation put forth by Wong [19] , Water Corporation [15] , and Barron, Wendling, Tucker, Green, Devkota and Donn [18] .

It is equally important to note that the irrigation in "The Green" is sourced from superficial aquifer, which is

Table 1. Water consumption (KL/household) in "The Green", Ridgewood, Butler, and Perth metropolitan region [32] [33] .

Areas\Years	2005	2006	2007	2008	2009	2010	2011	2012	2013	2014
Perth Metro*	375.0	387.5	362.5	362.5	352.5	347.5	344.3	331.5	336.6	347.5
Ridgewood	265.9	178.7	165.0	253.8	300.7	302.6	285.2	285.5	292.6	282.9
Butler (except "The Green")	258.7	282.6	371.1	391.0	364.2	346.2	333.6	290.9	269.5	278.3
The Green-Drinking Water				6.7	37.9	91.3	116.9	154.4	167.6	172.3
-Groundwater							54.1	188.8	167.5	115[Ψ]
-Total				6.7	37.9	91.3	171.0	343.3	335.1	287.3[Ψ]

*Household water consumption for Perth Metropolitan is calculated as a multiplication of average household size (2.6) with the average per-capita water consumption for each year. [Ψ]The allocated groundwater for residential irrigation in "The Green" is approximately 115 KL/household for 2013/14.

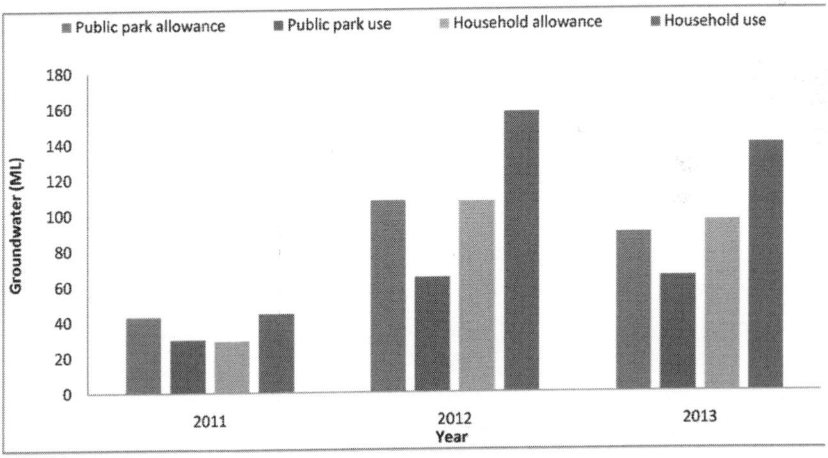

Figure 6. Total groundwater allocation and consumption in "The Green".

directly recharged by rainfall. The water sensitive land management practices implemented in "The Green" were reported to increase the aquifer recharge by 383 ML per year than that of pre-development conditions. It was also noticed that after abstracting 209 ML/year for groundwater irrigation network, still there would be an increase in net aquifer recharge by 174 ML/year [35] . Such an increase in aquifer recharge indicates that the groundwater would be a sustainable source for irrigating urban agriculture, vegetation, and recreational land- scaping as shown by previous studies of Western Australian Planning Commission [14] , and Nolasco [4] .

The irrigation trial has been operated for seven years but still approximately 20% of blocks are to be developed. Though new blocks are yet to be connected, the trend in water saving and irrigation efficiency is quite promising. Almost 50% less drinking water demand at residential settings and significant reduction in irrigation for agricultural activities in "The Green" than surrounding suburban areas and metropolitan average clearly indicate that the trialled system will achieve its target even at the full residential development of the suburb. However, an overconsumption of groundwater that was evident in "The Green" could be linked mainly to irrigation exemptions and other once-off usage. These transitional irrigation inconsistencies, as observed in Mawson Lakes Australia [17] , would gradually settle down after the suburban areas completely developed and irrigation system fully connected. Having said that, the complete analysis for the factors and parameters associated with the success of the irrigation trial requires completion of the residential development and longitudinal data on usage and behaviour towards water and irrigation systems. Joint effort of Urban Agriculturist, water providers and urban planners.

4.2. Inherent Socio-Economic Challenges

The community perceptions were found to be positive towards the efficient irrigation system and a majority of the consumers were reported to be satisfied with the system in "The Green". The higher level of satisfaction was mainly associated with the automatic irrigation operation, water savings, environmental benefits, and the trust on irrigation service providers. However, there were some social and economic challenges, mainly lack of clarity in institutional roles, and management of consumer expectations that need to be addressed for better management of potential alternative irrigation systems. This study shows that the fit-for-purpose irrigation with integrated land and water management brings the community members, government authorities, and service providers together to plan, implement and develop sustainable urban agricultural practices, and share mutual benefits.

4.3. Challenges in System Replication

The qualitative data analysis clearly shows that the majority of residents participated positively towards the efficient irrigation system for food production and agricultural activities at their home and community. Further, it shows an increasing awareness towards the efficient gardens and irrigation system in "The green" community. This is an important finding for the future development of urban agricultural activities with reliable efficient irrigation systems in similar location and climate across Australia.

Based on the findings of this study, alternative irrigation system sourced from rainwater has been successfully implemented in a neighbouring suburban area (Evermore Height, Alkimos). Successive new alternative irrigation projects in the Perth metropolitan and regional areas have also being formulated and developed taking Butler trial as a successful urban agriculture irrigation model [30] . These findings and successful replications clearly demonstrate that the fit-for-purpose irrigation system is instrumental in sustainable development of urban agriculture across Australia.

5. CONCLUSION

An alternative fit-for-purpose irrigation system was trialled in "The Green" community at Butler, Perth to understand sustainable water management for urban agriculture. The quantitative and qualitative data were collected and analyzed. The study suggested that a fit-for-purpose irrigation was a sustainable alternative for urban agriculture. Additionally, it offers significant irrigation efficiency in urban agriculture, and consequent reduction in water demand at urban residential settings. The water quality issues associated with fit-for-purpose irrigation appear to be nominal and do not affect the community acceptance. This indicates that the fit-for-purpose irrigation system with water efficient land management practices is highly encouraging for the sustainable development of urban agriculture and recreational landscaping across spatial levels.

ACKNOWLEDGEMENTS

Authors greatly acknowledge the key contribution of WA Water Corporation who provided relevant data and the fund to undertake this study. We thank Satterley Property Group for their financial contribution; Centre for Planning at Edith Cowan University for the operational management of the study. We are also grateful for Mr. Tim Perkins (Centre for Planning, ECU) and Dr. Bishnu Devkota (Infra Tech Pacific/WorleyParson, past-WA Water Corporation) for their generous comments on this paper.

REFERENCES

1. Mougeot, L.J.A. (2010) Agropolis: The Social, Political and Environmental Dimensions of Urban Agriculture. IDRC, London.

2. Zezza, A. and Tasciotti, L. (2010) Urban Agriculture, Poverty, and Food Security: Empirical Evidence from a Sample of Developing Countries. Food Policy, 35, 265-273.

3. Urban Agriculture Australia (2013) What Is Urban Agriculture and Why It Is Important? Urban Agriculture Australia, Canberra.

4. Nolasco, J. (2013) Sustainable Water Management for Urban Agriculture: Planting Justice. Pacific Institute, Oakland.

5. Lin, B.B., Philpott, S.M. and Jha, S. (2015) The Future of Urban Agriculture and Biodiversity-Ecosystem Services: Challenges and Next Steps. Basic and Applied Ecology, 16, 189-201.
http://dx.doi.org/10.1016/j.baae.2015.01.005

6. Lin, B.B. and Fuller, R.A. (2013) FORUM: Sharing or Sparing? How Should We Grow the World's Cities? Journal of Applied Ecology, 50, 1161-1168.

7. Lwasa, S., Mugagga, F., Wahab, B., Simon, D., Connors, J.P. and Griffith, C. (2015) A Meta-Analysis of Urban and Peri-Urban Agriculture and Forestry in Mediating Climate Change. Current Opinion in Environmental Sustainability, 13, 68-73.

8. Lovell, S.T. (2010) Multifunctional Urban Agriculture for Sustainable Land Use Planning in the United States. Sustainability, 2, 2499-2522.

9. Guendel, S. and Richards, W. (2002) Peri-Urban and Urban Livestock Keeping in East Africa—A Coping Strategy for the Poor, Scoping Study Commissioned by DFID. Natural Resources Institute, Aylesford.

10. Nugent, R. (2000) The Impact of Urban Agriculture on the Household and Local Economies. In: Bakker, N., Dubbeling, M., Gündel, S., Sabel-Koshella, U. and de Zeeuw, H., Eds., Growing Cities, Growing Food. Urban Agriculture on the Policy Agenda, Zentralstelle für Ern&aumlhrung und Landwirtschaft (ZEL), Feldafing, 67-95.

11. Mok, H.-F., Williamson, V., Grove, J., Burry, K., Barker, S.F. and Hamilton, A.J. (2014) Strawberry Fields Forever? Urban Agriculture in Developed Countries: A Review. Agronomy for Sustainable Development, 34, 21-43.

12. Cohen, N. and Reynolds, K. (2015) Resource Needs for a Socially Just and Sustainable Urban Agriculture System:

Lessons from New York City. Renewable Agriculture and Food Systems, 30, 103-114.

13. Water Corporation (2010) Perth Residential Water Use Study 2008/2009. Water Corporation, Perth.

14. Western Australian Planning Commission (2008) Better Urban Water Management. Western Australian Planning Commission, Perth.

15. Water Corporation (2009) Water Forever: Towards Climate Resilience. Water Corporation, Perth.

16. Smith, A., Pollock, D. and McFarlane, D. (2005) Opportunity for Additional Self Supply of Groundwater from the Superficial Aquifer beneath Metropolitan Perth. CSIRO: Water for a Healthy Country National Research Flagship, Canberra.

17. Hurlimann, A. (2008) Community Attitudes to Recycled Water Use: An Urban Australian Case Study Part 2. CRC for Water Quality and Treatment, Salisbury.

18. Barron, O., Wendling, L., Tucker, D., Green, M., Devkota, B. and Donn, M. (2010) Investigation of Techniques to Better Manage Western Australia's Non-Potable Water Resources. CSIRO: Water for a Healthy Country National Research Flagship, Canberra.

19. Wong, T.H. (2007) Water Sensitive Urban Design—The Journey Thus Far. Australian Journal of Water Resources, 10, 213-222.

20. Po, M., Nancarrow, B.E., Leviston, Z., Porter, N.B., Syme, G.J. and Kaercher, J.D. (2005) Predicting Community Behaviour in Relation to Wastewater Reuse: What Drives Decisions to Accept or Reject? CSIRO: Water for a Healthy Country National Research Flagship, Perth.

21. Nancarrow, B.E., Leviston, Z., Po, M., Porter, N.B. and Tucker, D.I. (2008) What Drives Communities' Decisions and Behaviours in the Reuse of Wastewater. Water Science and Technology, 57, 485-491.

22. Davis, C. and Farrelly, M. (2009) Demonstration Projects: Case Studies form Perth, Australia. National Urban Water Governance Program, Monash University, Melbourne.

23. Davis, C. and Farrelly, M. (2009) Demonstration Projects: Case Studies form Melbourne, Australia. National Urban Water Governance Program, Monash University, Melbourne.

24. Davis, C. and Farrelly, M. (2009) Demonstration Projects: Case Studies form South East Queensland, Australia. National Urban Water Governance Program, Monash University, Melbourne.

25. Department of Water (2010) Predicting the Future Demand for Water Resources in Western Australia. Department of Water, Perth.

26. Dzidic, P. and Green, M. (2012) Outdoing the Joneses: Understanding Community Acceptance of an Alternative Water Supply Scheme and Sustainable Urban Design. Landscape and Urban Planning, 105, 266-273.

27. Water Corporation (2005) Integrated Water Supply Scheme Source Development Plan 2005: Planning Horizon 2005-2050. Water Corporation, Perth.

28. Water Corporation (2007) Memorandum of Understanding: "The Green" at Brighton (Brighton Village 4) Superficial Groundwater Demonstration Project. Water Corporation, Perth.

29. Water Corporation (2007) Brighton Superficial Groundwater Supply Demonstration Project: Health Risk Management Plan. Water Corporation, Perth.

30. Satterley Property Group (2010) The End to the Waste of Precious Water Starts Here: Smartwater Initiative, The Green. Satterley Property Group, Perth.

31. Western Australian Planning Commission (2007) Liveable Neighbourhood: A Western Australian Government Sustainable Cities Initiatives. Western Australian Planning Commission, Perth.

32. Australian Bureau of Statistics (2013) 2011 Census Population and Housing: Greater Perth—Basic Community Profile (Catalogue Number 2001.0). Australian Bureau of Statistics, Canberra.

33. Water Corporation (2015) Reducing Water Use. Water Corporation, Perth.

34. Water Corporation (2015) Exemptions. Water Corporation, Perth.

35. GHD (2006) Satterley Property Group: Report on Brighton Stage 4. Hydrological Assessment. GHD, Perth.

3

ESTIMATION OF ENERGY BALANCE COMPONENTS OVER A DRIP-IRRIGATED OLIVE ORCHARD USING THERMAL AND MULTISPECTRAL CAMERAS PLACED ON A HELICOPTER-BASED UNMANNED AERIAL VEHICLE (UAV)

Samuel Ortega-Farías [1,*], Samuel Ortega-Salazar [2], Tomas Poblete [1], Ayse Kilic [2], Richard Allen [3], Carlos Poblete-Echeverría [4], Luis Ahumada-Orellana [1], Mauricio Zuñiga [1] and Daniel Sepúlveda [1]

[1]Centro de Investigación y Transferencia en Riego y Agroclimatología (CITRA) and Research program on Adaptation of Agriculture to Climate Change (A2C2), Universidad de Talca, Casilla 747, Talca 3460000, Chile

[2]School of Natural Resources and Civil Engineering, University of Nebraska-Lincoln, 311 Hardin Hall, 3310 Holdrege Street, Lincoln, NE 68588, USA

[3]Biological and Agricultural Engineering and Civil Engineering, Research and Ext. Center, University of Idaho, Kimberly, Moscow, ID 83844, USA

[4]Escuela de Agronomía, Pontificia Universidad Católica de Valparaíso, Quillota 2260000, Chile

ABSTRACT

A field experiment was carried out to implement a remote sensing energy balance (RSEB) algorithm for estimating the incoming solar radiation (Rsi), net radiation (Rn), sensible heat flux (H), soil heat flux (G) and latent heat flux (LE) over a drip-irrigated olive (cv. Arbequina) orchard located in the Pencahue Valley, Maule Region, Chile (35°25′S; 71°44′W; 90 m above sea level). For this study, a helicopter-based unmanned aerial vehicle (UAV) was equipped with multispectral and infrared thermal cameras to obtain simultaneously the normalized difference vegetation index (NDVI) and surface temperature ($T_{surface}$) at very high resolution (6 cm × 6 cm). Meteorological variables and surface energy balance components were measured at the time of the UAV overpass (near solar noon). The performance of the RSEB algorithm was evaluated using measurements of H and LE obtained from an eddy correlation system. In addition, estimated values of Rsi and Rn were compared with ground-truth measurements from a four-way net radiometer while those of G were compared with soil heat flux based on flux plates. Results indicated that RSEB algorithm estimated LE and H with errors of 7% and 5%, respectively. Values of the root mean squared error (RMSE) and mean absolute error (MAE) for LE were 50 and 43 W m^{-2} while those for H were 56 and 46 W m^{-2}, respectively. Finally, the RSEB algorithm computed Rsi, Rn and G with error less than 5% and with values of RMSE and MAE less than 38 W m^{-2}. Results demonstrated that multispectral and thermal cameras placed on an UAV could provide an excellent tool to evaluate the intra-orchard spatial variability of Rn, G, H, LE, NDVI and $T_{surface}$ over the tree canopy and soil surface between rows.

Keywords: evapotranspiration; irrigation; remote sensing; energy balance; olive orchard; UAV

1. INTRODUCTION

Olive orchards have increased significantly in Chile during the last decade, and water scarcity has become the main limitation to maintain olive oil production [1]. Studies indicate that Chile will face a significant reduction in rainfall of 20%–40% in the near future due to global climate change [2]. Under these conditions, sophisticated irrigation water management will be required to optimize water productivity (oil production per unit of consumed water, kg m^{-3}) and to maintain sufficient levels of yield and quality [3,4]. To achieve these targets, it is necessary to have a reliable method to quantify the olive water requirements and actual evapotranspiration (ETa). Traditional methods employ the ETo–crop coefficient approach (ETa = ETo × Kc, where ETo is the reference evapotranspiration (mm day^{-1}) and Kc is the crop coefficient) [5,6]. The accurate estimate of ETa or latent heat flux (LE) is an important factor for establishing irrigation strategies such as regulated deficit irrigation (RDI), which has been successfully applied on olive orchards to increase water productivity and olive oil quality [7,8,9,10]. In addition, Cohen et al., [11] and Ortega-Farias et al. [12] indicated that the site-specific irrigation management (SSIM) can be used as a technical tool to improve water productivity. In that case, SSIM characterizes the effect of intra- and inter-field spatial variability of soil and canopy vigor on the estimation of ETa and irrigation scheduling (irrigation frequency and timing).

Carrasco-Benavides et al. [13,14] and Samani et al. [15] have demonstrated that remote sensing energy balance (RSEB) models can be used to estimate the spatial variability of LE or ETa over sparse canopies such as orchards and vineyards. Several RSEB algorithms that vary in complexity compute ETa (mm h^{-1}) as residual from the energy balance (EB) equation [16,17,18,19,20,21,22,23]:

$$LE = Rn - H - G = ETa\frac{\lambda \rho_w}{3600} \qquad (1)$$

where LE is the latent heat flux (W m^{-2}); Rn is the net radiation (W m^{-2}); G is the soil heat flux (W m^{-2}); H is the sensible heat flux (W m^{-2}); λ is the latent heat of vaporization (J kg^{-1}); is the density of water (≅1000 kg m^{-3}); and 3600 converts from seconds to hours. Values of LE, Rn, G and H are estimated at the time of the satellite overpass.

RSEB algorithms have been found to be useful to account for the spatial and seasonal variability of ETa at regional and field scales when using satellite platforms like Landsat (8, 7 and 5) and ASTER [18,24,25]. Main limitations of current satellite platforms for practical application of RDI and SSIM, especially in heterogeneous canopies such as orchards and vineyards, are the lack of fine spatial resolution and real-time data at the field and sub-field scales [26,27]. Berni et al. [28,29] suggested that the two critical limitations for using current satellite sensors in real-time crop management are the lack of imagery having optimum spatial and spectral resolution and unfavorable revisit times for many crop stress-detection applications. In addition, Matese et al. [30] indicated that low-resolution images of traditional acquisition platforms fail to represent much of the intra-vineyard variability that is a key for application of the site-specific management (precision viticulture) of heterogeneous vineyards.

Canopy covers of commercial olive orchards are generally incomplete as a result of the canopy geometry that depends on canopy size, leaf area index, and plant density. The canopy training system and associated canopy geometry may affect significantly the partitioning of Rn into H, LE or G. Ortega-Farías and López-Olivari [31] found H generated at the soil surface to play a key role in the estimation of LE as a residual from EB equation for a drip-irrigated orchard having low values of leaf area index (LAI = 1.32 m^2m^{-2}), wetted area (Aw = 4.5%) and fractional cover (fr = 0.31). In regards to the orchard water requirements, good knowledge of the effect of intra-orchard spatial variability on the partitioning of Rn into H and G is necessary to improve the estimation of LE when using RSEB models. In addition, it is important to indicate that the current spatial resolutions of the thermal-based satellite platforms, which are 30 m or greater, introduce uncertainty to the quantification of LE or ET for heterogeneous canopies, where H generated at the soil surface between rows can be the primary component of the EB equation [1].

Modern optical remote sensors aboard an UAV provide the basis for the generation of very high spatial resolution images that can be used as inputs in the RSEB algorithms to estimate spatial variability of energy balance components. On this subject, multispectral and thermal sensors placed on an UAV could be a useful tool to advance our knowledge about the effect of intra-orchard spatial variability on the partitioning of Rn into G, H and LE over olive orchards whose row spacing range from 1 to 8 m. Thus, the main objective of this research was to implement a remote sensing energy balance (RSEB) algorithm to estimate the incoming solar radiation (Rsi), net radiation (Rn), sensible heat (H), soil heat (G) and latent heat (LE) over a drip-irrigated olive orchard using multispectral and thermal cameras placed on a helicopter-based UAV.

2. MATERIAL AND METHODS

2.1. Study Site Description

Flight campaigns collecting measurements of energy balance components and climatic variables were carried out during February and March 2014 on a flat and relatively uniform olive orchard located in the Pencahue Valley, Región del Maule, Chile (35°23′S; 71°44′W; 96 m. above sea level). The climate of the Pencahue Valley is a typical Mediterranean semiarid with a daily temperature equal to 14.8 °C from September to May. The historical average annual rainfall in the region is about 602 mm, falling mainly during the winter months (May to September). The summer period (December to February) is usually hot and dry

(3.5% of annual rainfall) while the spring period (September to November) is wet (16% of annual rainfall).

The experiment was conducted on a drip-irrigated commercial young olive orchard (cv. Arbequina) cultivated for oil production (Figure 1a). The olive trees were planted at high density (5.0 m between row and 1.5 m between trees) with north–south orientation and trained on a triangular hedgerow system (similar to a monocone system) with a tree height (h) of 3.2 m and canopy width (cw) of 1.55 m. The soil at the experimental site (Figure 1) is classified as Quepo series (Vertisol, family fine, thermic Xeric Apiaquerts) with a clay loam texture. The bulk density was 1.54 g cm^{-3} over the effective rooting depth of 0–60 cm, and volumetric water content at field capacity and wilting point were 0.33 and 0.19 m^3 m^{-3}, respectively.

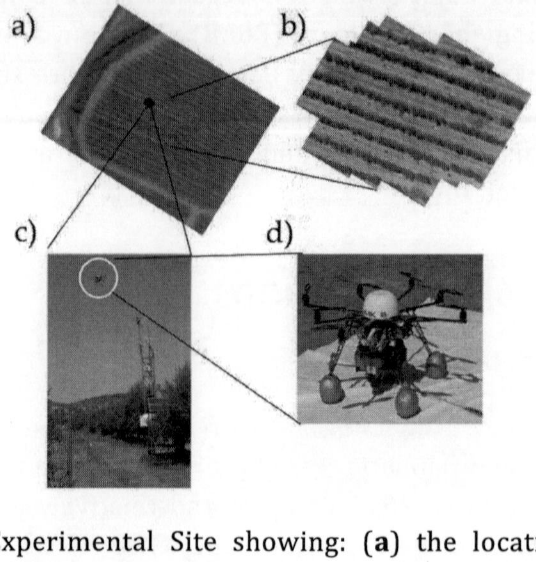

Figure 1. Experimental Site showing: (**a**) the location of eddy correlation (EC) system; (**b**) experimental site; (**c**) helicopter-based unmanned aerial vehicle (UAV) equipped with multispectral and infrared thermal cameras flying over an EC system installed over a drip-irrigated olive orchard; and (**d**) helicopter-based UAV.

To evaluate the impacts of irrigation management, the midday stem water potential (Ψx) was monitored at the time of the UAV overpass using a pressure chamber (PMS instruments, model 600, Albany, OR, USA). This was done using 12 fully expanded leaves wrapped in aluminum foil and encased in plastic bags for at least 2 h before the measurements [8]. The leaf area index (LAI) was measured weekly in 8 olive trees during the study periods using a plant canopy analyzer (LAI-2000, LI-COR, Lincoln, NE, USA) which was calibrated according to [31].

2.2. Measurements of Energy Balance and Climatic Data

A tower of 5.5 m high was installed to measure both surface energy balance components and climatic variables (Figure 1c). The latent heat flux (LE) and sensible heat flux (H) were measured using an eddy covariance system (EC) system oriented towards the predominant wind direction (south-east). The upwind fetch of the prevailing wind direction was about 500 m (for footprint analysis, see [1,31]). The EC consisted of a fast response open-path infrared gas analyzer (LI-7500 IRGA; LI-COR, Inc., Lincoln, NE, USA) and a three-dimensional sonic anemometer (CSAT3, Campbell Sci., Logan, UT, USA). LE and H measurements were made at 10 Hz, and means, standard deviations, and covariances were calculated over 15-min periods. Raw data of H and LE were post-processed considering corrections for air density [32], sonic temperature [33] and coordinate rotation [34]. During quality control, the energy balance closure was estimated using the ratio (Cr) of turbulent fluxes (H + LE) to available energy (Rn − G). When daily values of Cr were outside the range of 0.8 and 1.2, the entire day was excluded to reduce the uncertainty associated with errors in the LE and H measurements [35]. Assuming that the measurements of Rn and G were representative of the available energy in the olive orchard, the fluxes of H and LE for remaining data were forced to close the energy balance using the Bowen ratio approach [1,36,37,38]:

$$LEB = \frac{Rn - G}{1 + \beta} \quad (2)$$

$$HB = \frac{Rn - G}{1 + \beta^{-1}} \quad (3)$$

where β (=H/LE) is the Bowen ratio (dimensionless).

Net radiation (Rn), and incoming (Rsi) and outgoing (Rso) solar radiation were measured with a four-way net radiometer (CNR1, Kipp&Zonen Inc., Delft, The Netherlands). Wind speed (u) and wind direction (w) were monitored by a cup anemometer and a wind vane (03101-5, Young, MI, USA) and precipitation (Pp) was measured by a rain gauge (A730RAIN, Adcon Telemetry, Austria). Air temperature (T_a) and relative humidity (RH) were measured using a Vaisala probe (model HMP45C). In the tower, sensors of u, w, Pp, T_a, RH, Rn, Rsi, and Rso were installed at 5.5 m above the soil surface. A mobile tower with two CNR1 was also used to weekly measure canopy and soil albedos.

Soil heat flux (G) was estimated using eight flux plates installed on either side of the rows, with 4 sites placed inter-row and 4 placed within rows of trees. This arrangement took into account the shading effects by trees during the course of the day. The flux plates of constant thermal conductivity (HFT3, Campbell Sci., Logan, UT, USA) were placed at a 0.08 m depth. Two averaging thermocouple probes (TCAV, Campbell Sci., Logan, UT, USA) for measuring soil temperature (T_{soil}) were installed above each flux plate at depths of 0.02 and 0.06 m. Thermocouple probe signals were recorded at 15-min intervals on an electronic datalogger (CR3000, Campbell Sci, Logan, UT, USA). At each position, the soil heat flux was calculated by adding the measured flux at 0.08 m to the heat stored (S) in the layer above the heat flux plates [31]. All eight values for G were averaged during energy balance closure analysis.

2.3. Thermal and Multispectral Images Acquisition and Processing

For the drip-irrigated olive orchard, thermal and multispectral images were acquired from days of year (DOY) 58 to 80 using a helicopter-based UAV system equipped with multispectral and infrared thermal cameras (Figure 1d). The UAV system used eight propellers for propulsion and GPS-based, preprogrammed control. The cameras were controlled by an autopilot computer program to take photographs at user-selected waypoints to ensure complete coverage of the field. To minimize the shading effect of the cover, images were acquired around midday under clear sky conditions (Table 1). The flight height was configured at 60 m above ground level. This height allowed obtaining imagery at 3.3 cm and 6 cm spatial resolution in multispectral and thermal regions, respectively. A total of 10 images (dates) were available following filtering of images to remove those impacted by instrumental problems (based on Cr) and adverse meteorological conditions, especially with values of wind speed u >3.0 m s^{-1}.

Table 1. Meteorological conditions at the time of UAV overpass during the 2014 study period.

Flight Date (DOY)	Flight Time * (hh:mm)	Rsi ($W\ m^{-2}$)	T_a (°C)	RH (%)	u ($m\ s^{-1}$)	VPD (kPa)
58	12:46	901	24.7	37.3	1.1	1.2
59	12:08	892	22.9	38.4	1.5	1.1
62	12:09	856	20.9	54.2	1.9	1.3
64	12:05	833	22.8	50.0	1.5	1.4
65	11:56	876	22.4	38.5	1.5	1.0
66	12:08	857	23.1	39.4	2.7	1.1
69	12:08	879	23.1	28.4	2.1	0.8
73	12:38	839	21.1	50.8	1.8	1.3
74	12:49	834	22.3	46.5	1.4	1.3
80	12:40	801	18.1	40.8	1.0	0.9
	Mean	857 (±31)	22.1 (±1.8)	42.4 (±7.8)	1.6 (±0.5)	1.1 (±0.2)

* = local time (near solar noon); Rsi = incoming solar radiation; T_a = air temperature; RH = relative humidity; VPD = vapor pressure deficit; u = wind speed.

The multispectral camera (Mini MCA-6, Tetracam, Inc., Chatsworth, CA, USA) was configured with 6 individual sensors in the VIS–NIR spectral range at 530, 550, 570, 670, 700 and 800 nm, respectively. Image resolution was 1280 × 1024 pixels with 10-bit radiometric resolution and optics focal length of 9.6 mm. Images acquired with the Mini MCA-6 camera were preprocessed using the image handling application Pixel Wrench 2 (Tetracam Inc.) to produce multi-page TIF files. Radiometric calibration was performed using images of a spectralon white panel as a reference, which was acquired on the ground with the mini MCA-6 at the beginning and end of each flight [28]. The multispectral images were radiometrically corrected to convert the digital number of each pixel (brightness value) first into spectral radiance and then into reflectance as described by [39]. Afterwards, normalized difference vegetation index (NDVI) was computed by the traditional equation proposed by [40]:

$$\mathrm{NDVI} = \frac{\mathrm{NIR} - \mathrm{RED}}{\mathrm{NIR} + \mathrm{RED}} \tag{4}$$

where NIR and RED are the reflectances in the near infra-red (800 nm) and red (670 nm) bands, respectively. Maps were subsequently processed with a customized Matlab code (The Mathworks Inc., Natick, MA, USA) in order to estimate the intra-orchard spatial variability of NDVI.

The thermal infrared camera (EasIR-9, Wuhan Guide Infrared Co., Wuhan, China) has a resolution of 288 × 384 pixels and spectral response in the range 8–14 μm of the electromagnetic spectrum. The built-in sensor is a thermal detector (microbolometer) that avoids the requirement of an external cooling system. Thermal images were post-processed using the proprietary software IRAnalyser (Wuhan Guide Infrared Co., Ltd., Wuhan, China) with consideration given to specific parameters of emissivity, distance to the object, and measurements of T_a and RH at 5.5 above the soil surface (Figure 1c).

In each flight campaign, multispectral and thermal images were simultaneously taken at 20 points in the study area (Figure 1b). UAV images were acquired at nearly 0 nadir angle using a gimbal system. Only the central parts of the imagery were used to avoid border effects on the image-derived temperature [41]. The mosaicking of the images was performed using GPS ground control points which consisted on black and white panels that were easily identified from the images [42]. A Matlab code was developed to automatically identify the control points based on their respective reflectances. The fractional cover (f_r) was estimated assuming that the tree canopy presented values of NDVI > 0.40. The separation between the canopy and soil surface was performed manually by using discrimination and segmentation supervised techniques.

2.4. RSEB Algorithm Adapted for a Helicopter-Based UAV

A two-source algorithm was implemented to estimate the energy balance (EB) components above the drip-irrigated olive orchard using the thermal and multispectral data acquired by the miniaturized cameras onboard a helicopter-based UAV as input. The partitioning of instantaneous net radiation, which represented a mixture of canopy and soil was estimated as follows [43,44,45,46]:

$$\mathrm{Rne} = f_r \times \mathrm{Rn_C} + (1 - f_r) \times \mathrm{Rn_S} \quad (5)$$

where Rne is the estimated net radiation over the olive orchard; $\mathrm{Rn_C}$ and $\mathrm{Rn_S}$ are the contributions of the canopy and soil, respectively, to Rne; f_r is the fractional cover at nadir view. Rne, $\mathrm{Rn_C}$ and $\mathrm{Rn_S}$ are in W m^{-2}.

Instantaneous values for $\mathrm{Rn_C}$ and $\mathrm{Rn_S}$ were computed by establishing a balance between the long-wave and short-wave radiation, separately, for the soil surface and orchard canopy [16,43]:

$$Rn_c = (1 - \alpha_c)\, Rse + Lin - Lout_c - (1 - \varepsilon_c)\, Lin \quad (6)$$

$$Rn_s = (1 - \alpha_s)\, Rse + Lin - Lout_s - (1 - \varepsilon_s)\, Lin \quad (7)$$

where Rse is the instantaneous incoming shortwave radiation; α_s and α_c are soil and canopy albedos, respectively; Lin is the instantaneous incoming longwave radiation; $Lout_c$ and $Lout_s$ are the instantaneous outgoing longwave radiations from the soil and canopy, respectively; and ε_c and ε_s are the surface thermal emissivities (dimensionless) from the soil and canopy, respectively. In this study, instantaneous values of Rse were estimated according to [16]. Rse, Lin, $Lout_c$ and $Lout_s$ are in W m^{-2}.

Values of Lin, $Lout_c$ and $Lout_s$ at the time of UAV overpass were computed as [16,43]:

$$Lin = \varepsilon_{atm} \sigma T_a^4 \quad (8)$$

$$\varepsilon_{atm} = 0.85 \left(-\ln\left(\tau_{sw}\right)\right)^{0.09} \quad (9)$$

$$Lout_c = \varepsilon_c \sigma T_c^4 \quad (10)$$

$$Lout_s = \varepsilon_s \sigma T_s^4 \quad (11)$$

where σ is the Stefan-Boltzmann constant (W m^{-2} K^{-4}); T_a, T_c and T_s are the air, canopy and soil temperatures between rows, respectively; is the atmospheric transmissivity (dimensionless); and ε_{atm} is the atmospheric emissivity (dimensionless). Values of $\varepsilon_c$and ε_s were assumed as 0.98 and 0.95, respectively [29,47]. Values of T_a, T_c and T_s are in °K. Equation (9) follows the methodology indicated by [16]

The soil and canopy contributions to total sensible heat were estimated using the following equations [44,45]:

$$He = f_r \times H_c + (1 - f_r) \times H_s \tag{12}$$

$$H_c = \frac{\rho_a C_p (T_c - T_a)}{r_a^c + r_a^a} \tag{13}$$

$$H_s = \frac{\rho_a C_p (T_s - T_a)}{r_a^s + r_a^a} \tag{14}$$

where He, H_c and H_s are the sensible heat fluxes (W m^{-2}) above the orchard, tree canopy and soil surface, respectively; ρ_a is the air density (kg m^{-3}); C_p is the specific heat of dry air $(\approx 1005\ J\ kg^{-1}\ {}^\circ K^{-1})$; is the bulk boundary layer resistance of the vegetative elements in the canopy (s m^{-1}); is the aerodynamic resistance between canopy source height and reference level (s m^{-1}); and is the aerodynamic resistance between the soil and canopy source height (s m^{-1}). The estimation of , and are estimated according to [35,48] (see Appendix A).

Finally, instantaneous values of latent heat flux over the olive orchard were computed as follows:

$$LEe = Rne - He - Ge \tag{15}$$

In this case, estimated heat flux (Ge, W m^{-2}) was estimated as:

$$Ge = 0.3236 \times Rne - 51.52 \tag{16}$$

Equation (16) was developed using a database collected for an olive orchard during the 2009–2010 and 2010–2011 growing seasons [1,49]. The measured values of α_s, α_c, and LAI ranged between 0.15–16, 0.11–0.12 and 1.15–1.29 $m^2\ m^{-2}$ while estimated values of f_r using the UAV ranged between 0.25 and 0.28.

2.5. Statistical Analysis

An analysis of spatial variability of surface temperature ($T_{surface} = T_s + T_c$), NDVI and energy balance components was made above the tree canopy and above the soil surface between rows. Due to the high resolution of the thermal and multispectral images, crown edges were excluded to ensure that only pure vegetation and soil pixels were used in the comparison between estimated and observed fluxes [50]. In this case, Rne, Ge, He, and LEe were computed by averaging values of 405 × 405 (6 cm × 6 cm) pixels that fell within the footprint of the EC.

The validation of sub-models that compute Rne, Ge, He and LEe was carried out using the ratio of estimated to observed values (b), mean absolute error (MAE), root-mean-square error (RMSE) and index of agreement (Ia) [51,52]. The student's *t*-test analysis was applied to check whether b was significantly different from unity at the 95% confidence level. In addition, the sub-model for estimating incoming solar radiation (Rse) was included in the validation. For the validation, measurements of Rsi, Rn, G, HB and LEB between 12:00 and 13:00 were linearly interpolated to compare with the estimated values (Rse, Rne, Ge, He and LEe) at the time of UAV overpass (Table 1).

3. RESULTS

Meteorological data obtained at the time of UAV overpasses are indicated in Table 1. Dry and hot atmospheric conditions were observed during the campaigns where values of Rsi, T_a, RH, u and vapor pressure deficit (VPD) ranged between 801–901 W m^{-2}, 18.1–24.7 °C, 28.4%–54.2%, 1.0–2.7 m s^{-1}, and 0.80–1.4 kPa, respectively. Within these atmospheric environments, values of $\Psi_x$ranged between −1.5 and −2.1 MPa indicating that the drip-irrigated olive orchard was maintained under non and moderate water stress conditions during the study period [8,53].

The accuracy of the EC measurements above the olive orchard was checked at 15 min time intervals using energy balance closure which presented a coefficient of determination of 0.92 and Cr of 0.86 (Figure 2). Table 2 shows that the energy balance closure at the time of UAV overpass was 0.86 (±0.03) indicating that the orchard energy balance was systematically imbalanced by about 14%. Several researchers have indicated that turbulent fluxes using the EC technique were less than available energy for olive orchard with EC imbalances between 5% and 26% [31,36,37,54,55,56,57]. Lack of closure may be associated to errors in the measurements of Rn and G [58,59,60], energy storage within the olive tree biomass [61], low wind speed [62] and heterogeneity of surface energy balance of the region [63]. Finally, [53,64] suggested that EC uncertainties can be corrected using the Bowen-ratio approach (β = H/LE).

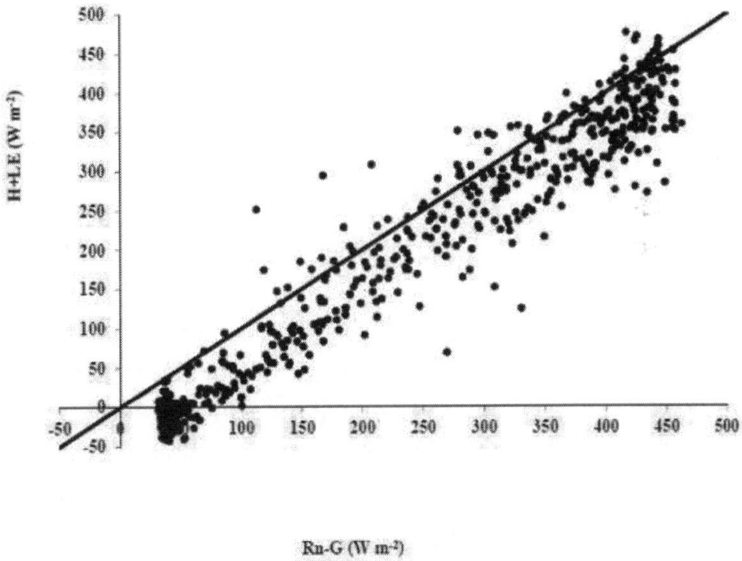

Figure 2. A regression analysis between turbulent energy fluxes (H + LE) from the eddy correlation system and available energy (Rn − G) over a drip-irrigated olive orchard. The solid line represents the 1:1 line.

Table 2. Energy balance closure (Cr), Bowen ratio (β) and instantaneous ratios of latent (LEB), sensible (HB) and soil (G) heat fluxes to net radiation (Rn) over a drip-irrigated olive orchard at the time of UAV overpasses. Instantaneous ratios of Rn to incoming solar radiation (Rsi) are included.

DOY	C_r	β	Rn/Rsi	LEB/Rn	HB/Rn	G/Rn
58	0.88	3.08	0.67	0.18	0.56	0.25
59	0.87	3.87	0.67	0.15	0.59	0.25
62	0.90	3.57	0.67	0.16	0.59	0.25
64	0.81	2.61	0.65	0.21	0.55	0.24
65	0.86	2.61	0.67	0.21	0.54	0.25
66	0.85	2.62	0.67	0.21	0.54	0.25
69	0.87	4.11	0.66	0.15	0.60	0.25
73	0.81	2.77	0.66	0.20	0.55	0.25
74	0.86	2.88	0.66	0.19	0.56	0.25
80	0.88	3.30	0.67	0.18	0.58	0.24
Average	0.86 (±0.03)	3.09 (±1.7)	0.66 (±0.01)	0.18 (±0.02)	0.57 (±0.02)	0.25 (±0.00)

At the time of UAV overpass (near solar noon), Table 2 indicates that total bulk β above the soil-canopy system varied between 2.61 and 4.11 with maximum values observed on DOY 69. Mean values of LE accounted for 15%–21% of Rn while those of H were between 54% and 60% of Rn. The analysis indicated that the main component of the surface energy balance over the drip-irrigated olive orchard was H. Rn/Rsi and G/Rn ratios were very stable during the study period with mean values of 0.66 and 0.25, respectively. Results are similar to those found by López-Olivari et al. [1] who indicated that daytime values of LE, H and G over a drip-irrigated olive orchard (f_r = 30%) were between 28%–47%, 34%–68% and 2%–6% of Rn, respectively, while ratios of transpiration and soil evaporation to ETa ranged between 0.64–0.74 and 0.26–0.36, respectively.

The intra-orchard spatial variability of NDVI and surface temperature is indicated in Figure 3 based on multispectral and thermal imagery collected by the helicopter-based UAV. This Figure confirms the degree that $NDVI_c$ (canopy) was larger than $NDVI_s$ (soil surface) and that T_c was less than T_s during the study period. Figure 4 indicates that averaged values of $NDVI_c$ and $NDVI_s$ during the data collection period ranged between 0.63–0.71

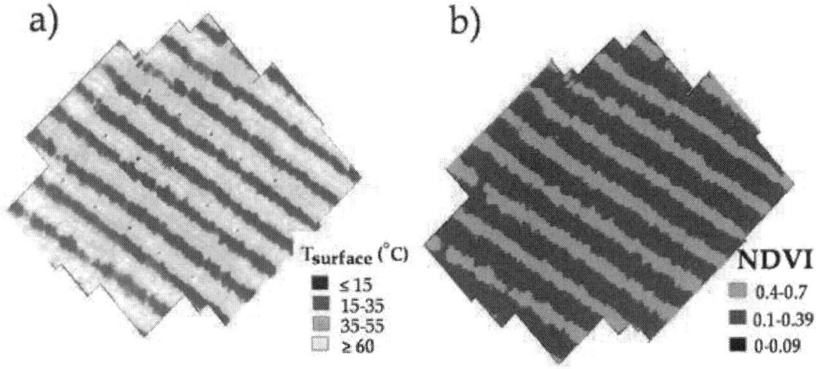

Figure 3. (**a**,**b**) Averaged spatial distribution of the normalized difference vegetation index (NDVI) and surface temperature (Tsurface) over a drip-irrigated olive orchard using multispectral and thermal cameras, respectively, placed aboard an unmanned aerial vehicle (UAV).

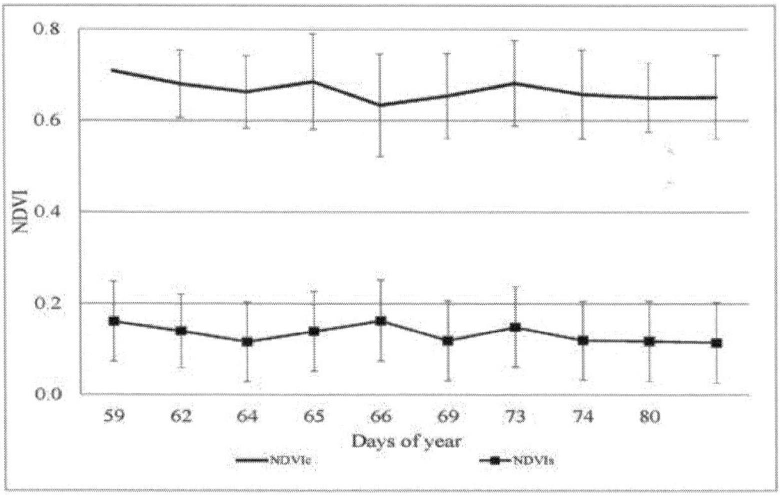

Figure 4. Mean values of the normalized difference vegetation index of (NDVI) of a drip-irrigated olive orchard using a multispectral camera placed aboard an unmanned aerial vehicle (UAV). Sub-indexes "c" and "s" denote values from the canopy and soil surface, respectively. Vertical lines indicate one standard deviation.

and 0.11–0.16, respectively while Figure 5 shows that T_c and T_s were between 16.1–27.7 and 39.9–51.9 °C, respectively. The coefficients of variation (CV) of $NDVI_c$, $NDVI_s$, T_c and T_s were 5.1%, 6.2%, 8.0% and 9.3%, respectively. These results indicate that spatial variability of T_c and T_s was greater than that of $NDVI_c$ and $NDVI_s$ during the flight campaigns. Furthermore, $NDVI_c$, $NDVI_s$, T_c and T_s presented a low spatial variability because CV values were less than 15% [65].

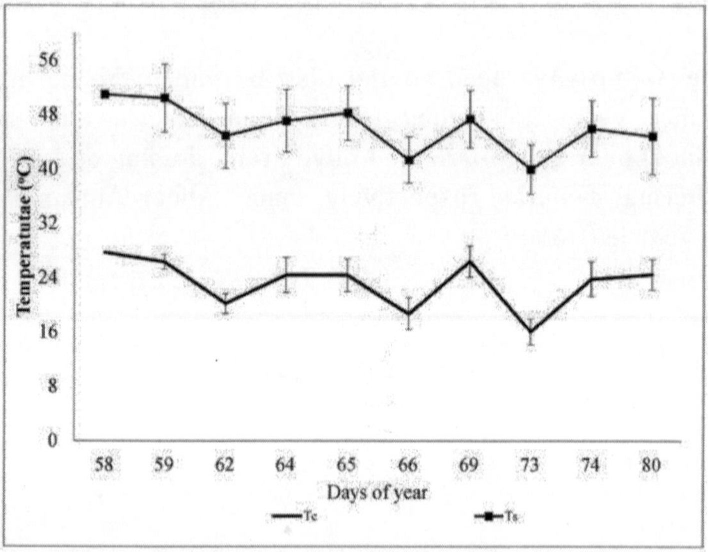

Figure 5. Mean values of surface temperature over a drip-irrigated olive orchard using a thermal camera placed aboard an unmanned aerial vehicle (UAV). Sub-indexes "c" and "s" denote values from the canopy and soil surface, respectively. Vertical lines indicate one standard deviation.

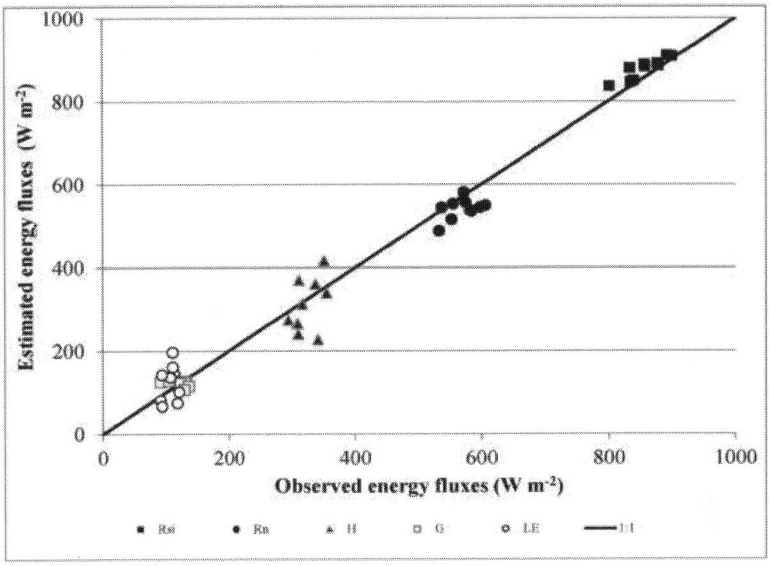

Figure 6. Comparisons at the time of UAV overpass between observed (axis X) and estimated (axis Y) values of bulk (canopy and soil) incoming solar radiation (Rsi) net radiation (Rn), soil heat flux (G), sensible heat flux (H), and latent heat flux (LE) over a drip irrigated olive orchard.

Figure 6 shows comparisons between observed and estimated values of solar radiation and energy balance components at the time of UAV overpass (near solar noon). For solar radiation (black squares) and net radiation (black circles), all points were close to the 1:1 line. Values of RMSE and MAE for solar radiation were 24 and 21 W m^{-2}, respectively, while those for net radiation were 38 and 33 W m^{-2} (Table 3). The *t*-test indicated that the Rse/Rso ratio was not significantly different from unity, indicating that observed and estimated values were similar. However, the Rne/Rno ratio was statistically less than unity, where net radiation was underestimated by about 5.0%. These results are similar to those indicated in the literature for sparse canopies when using satellite-based remote sensing (SBRS) models. For a heterogeneous landscape in China, Liu et al. [66] showed values of

RMSE and MAE equal to 51 and 25 W m^{-2}, respectively, when using a simplified SBRS model. For a drip-irrigated vineyard, Carrasco-Benavides et al. [14] indicated that the METRIC model overestimated instantaneous Rn by about 11% with RMSE = 69.4 W m^{-2} and MAE = 63 W m^{-2}. For a drip-irrigated olive orchard, Ortega-Farias et al. [67] indicated that the METRIC model underestimated the instantaneous values of Rn by about 3% with RMSE and MAE values of 40 and 33 W m^{-2}, respectively.

Table 3. Validation of sub-models that compute solar radiation (Rsi), net radiation (Rn), sensible heat flux (H), soil heat flux (G) and latent heat flux (LE) over a drip-irrigated olive orchard at the time of UAV overpass.

Variable	RMSE (W m^{-2})	MAE (W m^{-2})	b	Ia	*t*-Test
Rsi	24	21	1.02	0.92	T
Rn	38	33	0.95	0.88	F
G	19	16	1.02	0.66	T
HB	56	46	0.95	0.74	F
LEB	50	43	1.07	0.54	F

RMSE = root mean square error; MAE = mean absolute error; b = ratio of observed to computed values; Ia = index of agreement; T = null hypothesis (b = 1) True; F = alternative hypothesis (b ≠ 1).

Comparisons between measured and estimated sensible heat fluxes at the time of UAV overpasses (Figure 6—black triangles) show that the points were evenly distributed along the 1:1 line. In this case, the sub-model underestimated the instantaneous sensible heat fluxes by about 5% with RMSE and MAE equal to 56 and 46 W m^{-2}, respectively (Table 3). The best agreement between measured and estimated sensible heat fluxes was observed on DOY 65 (5 March) where the differences between HB and He were less than 20 W m^{-2}. The greatest disagreement was observed on DOY 69 where HB and He were 351 and 416 W m^{-2}, respectively. In general, largest differences were observed when wind speed was >2.7 m s^{-1}. Over heterogeneous surfaces, Chávez et al. [64] indicated that the RMSE was 23 W m^{-2} when a simplified SBRS model was applied to estimate the sensible heat flux at the time of satellite overpass. For a drip-irrigated vineyard, Carrasco-Benavides et al. [14] found that METRIC with the calibrated functions of LAI, roughness length ( ) and G overestimated instantaneous values of H by about 16% with RMSE and MAE equal to 67 and 57 W m^{-2}, respectively.

The comparison between measured and modeled soil heat fluxes (open squares) in Figure 6 shows that there was an equilibrated distribution of points around the 1:1 line. In this case, instantaneous values of G were estimated with an error = 2%, RMSE = 19 W m^{-2} and MAE = 16 W m^{-2} (Table 3). Over heterogeneous surfaces, Shaomin et al. [68], using a remote sensing model, estimated soil heat fluxes with an error = 19% and RMSE = 23 W m^{-2} and suggested that errors could be attributed to differences in spatial/temporal scales between satellite data and ground-truth measurements. Similarly, Carrasco-Benavides et al. [14] observed for a drip-irrigated vineyard that METRIC simulated G with an error, RMSE and MAE of 17%, 34 W m^{-2} and 21 W m^{-2}, respectively.

For the instantaneous latent heat flux, Figure 6 (open circles) shows that the points were grouped above the 1:1 line with a RMSE = 50 W m^{-2} and MAE = 43 W m^{-2} (Table 3). The LEe/LEo ratio was significantly different from unity suggesting that the RSEB algorithm overestimated instantaneous values of LE by about 7%. For drip-irrigated vineyards, González-Dugo et al. [69] observed errors of 18% in the estimation of LE when using a two-source energy balance (TSEB) model and Landsat scenes. For a drip-irrigated Merlot vineyard, Carrasco-Benavides et al. [14] indicated that METRIC using calibrated functions, overestimated LE with error = 17%, RMSE = 60 W m^{-2} and MAE = 48 W m^{-2}. For a drip-irrigated olive orchard, Ortega-Farias et al. [70] observed that the METRIC model underestimated LE by about 11% with RMSE = 29 W m^{-2} and MAE = 26 W m^{-2}.

For high-resolution multispectral and thermal imagery obtained by aircraft mounted sensors, Xia et al. [71] observed that a two-source energy balance (TSEB) was able to simulate the energy balance components over two vineyards with MAE values ranging between 15 and 90 W m^{-2}. These authors concluded that the TSEB model is fairly robust and able to derive reliable LE and ET patterns at sub-field scale under a wide range of environmental conditions. For a lightweight thermal camera mounted on a UAV, Homann et al. [72] indicated that the TSEB estimated the energy balance components over a barley field with RMSE and MAE ranging between 38–94 and 33–84 W m^{-2}, respectively. They suggested that the lightweight thermal camera placed on UAV platform provided high spatial and temporal resolution data for estimating energy balance fluxes over a barley field.

At the time of the UAV overpass, Figure 7 shows maps with the intra-orchard spatial variability of the energy balance components estimated by the RSEB algorithm. The Figure also shows that the maps at very high spatial resolution (pixels = 6 cm × 6 cm) are detailed enough to show differences between the energy balance components above the tree canopy and the soil surface between rows. Figure 7a indicates that Rn_c was greater than Rn_s with CV values of 5.5% and 6.1%, respectively. The average values of H_c and H_s were 46 and 397 W m^{-2} while those of LE_c and LE_s were 573 and 30 W m^{-2}, respectively. In addition, the values of CV ranged between 6% and 13% for H_c, H_s, LE_c and LE_s suggesting that the spatial variability of turbulent fluxes was low for the olive orchard [65].

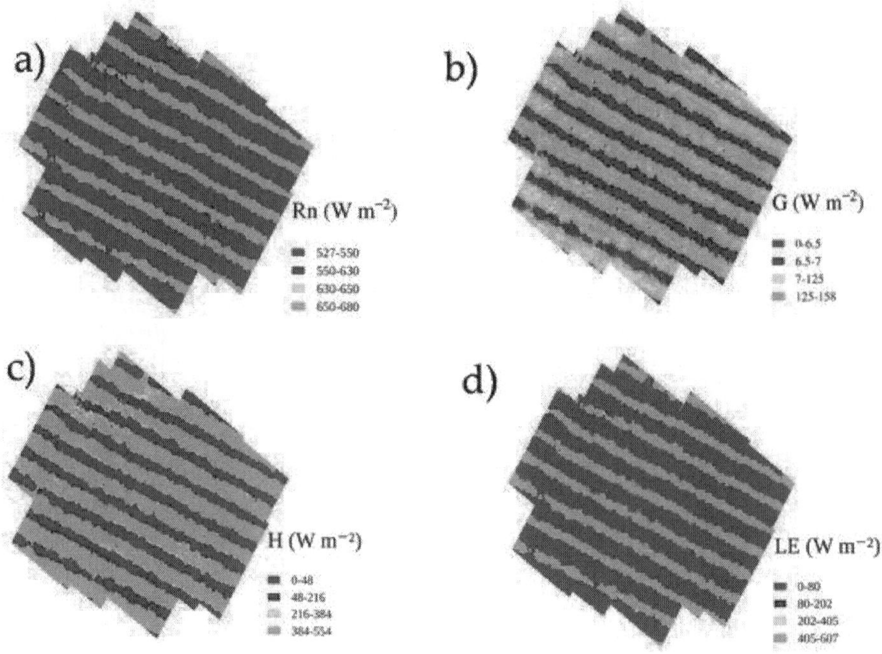

Figure 7. Average spatial distribution of estimated values of (**a**) net radiation (Rn); (**b**) soil heat flux (G); (**c**) sensible heat flux (H); and (**d**) latent heat flux (LE) over the olive canopy and soil surface between rows using a thermal sensor placed aboard an unmanned aerial vehicle (UAV).

4. DISCUSSIONS

The results (RMSE, MAE and errors) obtained in this research coincide with those reported in the literature for sparse vineyards and orchards using satellite, aircraft and UAV platforms [14,64,66,67,68,69,71,72]. The implementation of the RSEB algorithm was well defined and the only input data used to simulate orchard energy balance were measurements of climatic variables (Ta, u, and RHa) and UAV images (NDVI and $T_{surface}$). The results also indicated that the maps at very high spatial resolution were able to show significant differences between the energy balance fluxes above the tree canopy and the soil surface between rows. Moreover, the values of Rn_c and LE_c were greater than those of Rn_s and LE_s while those of T_c and H_c were significantly less than those of T_s and H_s (Figure 7). This study demonstrated that multispectral and thermal cameras placed on an UAV could provide an excellent tool to evaluate the effects of spatial variability on the partitioning of Rn into G, H and LE over heterogeneous orchards. Finally, UAV could be a useful complement to current satellite platforms for estimating intra-field spatial variability of the energy balance components in order to improve the estimation of water requirements of sparse or complex canopies such as orchards and vineyards which, in turn, present different plant densities and fractional covers.

The water requirements of drip-irrigated olive orchards require a good knowledge of H/Rn, G/Rn and LE/Rn ratios, which depends on tree vigor. The vigor is manifested through the canopy size, LAI, and f_r. In this regard, López-Olivari et al. [1] suggested that the canopy architecture (tree density, canopy size, LAI, and f_r), pruning, and irrigation systems (percentage of wetted area of soil surface) have an important effect on the partitioning of Rn into LE, H and G. Similarly, Testi et al. [62] suggested that the energy exchange between the soil surface and the tree canopy is strongly affected by the orchard architecture and fractional cover. According to the canopy structure and fractional cover, H generated at the soil surface can be a major contributor to the orchard energy balance, which plays a key role in the tree transpiration and evaporation. During this study, the canopy cover was generally incomplete as a result of the regimented structure imposed by the triangular hedgerow system. This training system consisted of widely spaced trees that allowed for deep penetration of sunlight and air turbulence into the canopy. For the drip-irrigated olive orchard, a constant shape of the canopy was maintained with values of LAI and f_r ranging between 1.15–1.29 $m^2\ m^{-2}$ and 0.25–0.28 $m^2\ m^{-2}$, respectively. The soil surface between rows was kept free of weeds (cover crops) during February and March; during this time rainfall was less than 5 mm. In addition, the soil surface (0–20 cm) between rows was very dry and the soil surface wetted by drip emitters was 4.5%. In these orchard conditions, values of α_s, α_c, $NDVI_c$ and $NDVI_s$ were quite constant during the study period, making it possible to carry out a good comparison between the observed and the estimated values of the net radiation and soil heat flux at the time of the UAV overpass. In addition, sensible heat flux generated at the soil surface between rows was the main component of the orchard energy balance which affected the partitioning of LE into LE_c (transpiration) and LE_s (soil evaporation). In this case, the H_c/H_s and LE_c/LE_s ratios were 0.21 and 7.64, respectively, for the olive orchard that was also maintained under non and moderate water stress conditions. Finally, $NDVI_c$, $NDVI_s$, T_c and T_s presented a low spatial variability (CV < 15%) because LAI and f_r were maintained constant by the triangular hedgerow system.

5. CONCLUSIONS

This study indicates that the use of both multispectral and thermal cameras onboard of the UAV allowed acquisition of high-resolution images to assess the intra-field spatial variability of energy balance components over a drip-irrigated olive orchard under non and moderate water stress conditions (midday Ψx > −2.1 MPa). Simulated net radiation, soil, sensible, and latent heat fluxes were generally in good agreement with flux measurements from an eddy covariance system located within the olive orchard. In this experiment, RSEB algorithm estimated instantaneous values of Rn, H, G and LE with errors less than 7%. Disagreement between modeled and measurements were associated with the estimation of sensible heat flux, especially on days where wind speed was greater than >2.7 m s^{-1}.

In addition, this experiment demonstrated that multispectral and thermal cameras placed on an UAV could provide an excellent tool to evaluate the effects of intra-orchard spatial variability of Rn, G, H, LE, NDVI and $T_{surface}$ over the tree canopy and soil surface between rows. In this cases, the maps at very high spatial resolution (pixels = 6 cm × 6 cm) are detailed enough to show differences between the energy balance components above the tree canopy and the soil surface between rows. In addition, sensible heat flux generated at the soil surface between rows was the main component of the orchard energy balance, which affected the partitioning of LE into LEc (transpiration) and LEs (soil evaporation). Finally, the coefficient of variation indicated that intra-canopy and intra-row spacing spatial variability of $T_{surface}$ was greater than that of NDVI during the flight campaigns.

Future improvements will incorporate spatially distributed multispectral data into the RSEB models in order to estimate the intra-field spatial variability of albedo, leaf area index, and surface emissivity. In addition, parameterization of aerodynamic resistance approaches will be required to simulate sensible heat flux over heterogeneous orchards, which consist of widely spaced trees that allow for deep penetration of sunlight and air turbulence into the canopy. Finally, future work will evaluate the parameterization of turbulent fluxes and available energy over the tree canopy and soil surface between rows.

ACKNOWLEDGMENTS

This study was supported by the Chilean government through the projects FONDECYT (No. 1130729) and FONDEF (No. D10I1157). The authors would like to thank Manuel Barrera and Alvaro Ried from the "Olivares de Quepu" Company for their technical support.

AUTHOR CONTRIBUTIONS

Samuel Ortega-Farias conceived and designed the experiments, supported the statistical analysis, and wrote the original manuscript draft. Ayse Kilic, Richard Allen and Carlos Poblete-Echeverría contributed extensively to results interpretation and discussions and editing of the manuscript. Luis Ahumada-Orellana and Mauricio Zuñiga supported the field experiment and helped with editorial contributions. Samuel Ortega-Salazar and Tomas Poblete analyzed UAV images and supported the statistical analysis. Daniel Sepúlveda performed the UAV flights and analyzed field data. All co-authors contributed valuable information to the final version.

CONFLICTS OF INTEREST

The authors declare no conflict of interest.

REFERENCES

1. López-Olivari, R.; Ortega-Farías, S.; Poblete-Echeverría, C. Partitionning of net radiation and evapotranspiration over a superintensive drip-irrigated olive orchard. Irrig. Sci. 2016, 1, 17–31.

2. Intergovernmental Panel on Climate Change (IPCC). The Fifth Assessment Report (AR5); World Meteorological Organization (WMO): Geneva, Switzerland; United Nations Environment Programme (UNEP): Nairobi, Kenya, 2013.

3. Fereres, E.; Soriano, M.A. Deficit irrigation for reducing agricultural water use. J. Exp. Bot. 2007, 58, 147–159.

4. Ortega-Farias, S.; Irmak, S.; Cuenca, R.H. Special issue on evapotranspiration measurement and modeling. Irrig. Sci. 2009,28, 1–3.

5. Allen, R.G.; Pereira, L.S.; Raes, D.; Smith, M. Crop Evapotranspiration. Guidelines for Computing Crop Water Requirements; FAO Irrigation and Drainage Paper No. 56; Food and Agriculture Organization of the United Nations: Rome, Italy, 1998.

6. Ortega-Farías, S.; Cuenca, R.H.; English, M. Hourly grass evapotranspiration in modified maritime environment. J. Irrig. Drain. Syst. 1995, 6, 369–373.

7. Alegre, S.; Marsal, J.; Mata, M.; Arbones, A.; Girona, J.; Tovar, M. Regulated deficit irrigation in olive trees (Olea europaeaL. cv. 'Arberquina') for oil production. Acta Hortic. 2002, 586, 259–262.

8. Flores, F.; Ortega-Farías, S. Effect of Three Levels of Water Application on Oil Yield and Quality for an Olive ('Picual') Orchard. Acta Hortic. (ISHS) 2011, 889, 317–322.

9. Patumi, M.; D'Andria, R.; Marsilio, V.; Fontanazza, G.; Morelli, G.; Lanza, B. Olive and olive oil quality after intensive monocone olive growing (Olea europaea L., cv. Kalamata) in different irrigation regimes. Food Chem. 2002, 77, 27–34.

10. Tognetti, R.; d'Andria, R.; Morelli, G.; Alvino, A. The effect of deficit irrigation on seasonal variations of plant water use inOlea europaea L. Plant Soil 2005, 273, 139.

11. Cohen, Y.; Alchanatis, V.; Meron, M.; Saranga, S.; Tsipris, J. Estimation of leaf water potential by thermal imagery and spatial analysis. J. Exp. Bot. 2005, 56, 1843–1852.

12. Ortega-Farias, S.; Rigetti, T.; Acevedo, C.; Matus, F.; Moreno, Y. Irrigation-management decision system (IMDS) for vineyards (Region VI and VII of Chile). Integrated Soil Water Management for Orchard Development. FAO Land Water Bull.2005, 10, 59–64.

13. Carrasco-Benavides, M.; Ortega-Farías, S.; Lagos, L.O.; Kleissl, J.; Morales, L.; Poblete-Echeverría, C.; Allen, R.G. Crop coefficients and actual evapotranspiration for a drip-irrigated Merlot vineyard using multispectral satellite images. Irrig. Sci.2012, 30, 537–553.

14. Carrasco-Benavides, M.; Ortega-Farías, S.; Lagos, L.O.; Kleissl, J.; Morales-Salinas, L.; Kilic, A. Parameterization of the

Satellite-Based Model (METRIC) for the estimation of instantaneous surface energy balance components over a drip-irrigated vineyard. Remote Sens. 2014, 6, 11342–11371.

15. Samani, Z.; Bawazir, A.S.; Bleiweiss, M.; Skaggs, R.; Longworth, J.; Tran, V.D.; Pinon, A. Using remote sensing to evaluate the spatial variability of evapotranspiration and crop coefficient in the lower Rio Grande Valley, New Mexico. Irrig. Sci.2009, 1, 93–100.

16. Allen, R.G.; Tasumi, M.; Morse, A.; Trezza, R.; Wright, J.L.; Bastiaanssen, W.; Kramber, W.; Lorite, I.; Robison, C.W. Satellite-based energy balance for mapping evapotranspiration with internalized calibration (METRIC) applications. J. Irrig. Drain. Eng. ASCE 2007, 4, 395–406.

17. Bastiaanssen, W.G.M.; Menenti, M.; Feddes, R.A.; Holtslag, A.A.M. A remote sensing surface energy balance algorithm for land (SEBAL)—1. Formulation. J. Hydrol. 1998, 1–4, 198–212.

18. Bastiaanssen, W.G.; Noordman, J.M.; Pelgrum, H.; Davids, G.; Thoreson, B.P.; Allen, R.G. SEBAL model with remotely sensed data to improve water-resources management under actual field conditions. J. Irrig. Drain. Eng. ASCE 2005, 1, 85–93.

19. Bastiaanssen, W.G.M. SEBAL-based sensible and latent heat fluxes in the irrigated Gediz Basin, Turkey. J. Hydrol. 2000,229, 87–100.

20. Tasumi, M.; Allen, R.G.; Trezza, R.; Wright, J.L. Satellite-Based energy balance to assess within-population variance of crop coefficient curves. J. Irrig. Drain. Eng. ASCE 2005, 1, 94–109.

21. Tasumi, M.; Allen, R.G.; Trezza, R. At-surface reflectance and albedo from satellite for operational calculation of land surface energy balance. J. Hydrol. Eng. 2008, 2, 51–63.

22. Teixeira, A.; Bastiaanssen, W.G.M.; Ahmad, M.D.; Bos, M.G. Reviewing SEBAL input parameters for assessing evapotranspiration and water productivity for the Low-Middle Sao Francisco River basin, Brazil Part A: Calibration and validation. Agric. For. Meteorol. 2009, 3–4, 462–476.

23. Allen, R.; Irmak, A.; Trezza, R.; Hendrickx, J.M.H.; Bastiaanssen, W.; Kjaersgaard, J. Satellite-based ET estimation in agriculture using SEBAL and METRIC. Hydrol. Process. 2011, 26, 4011–4027.

24. Gowda, P.H.; Chávez, J.L.; Colaizzi, P.D.; Evett, S.R.; Howell, T.A.; Tolk, J.A. Remote sensing based energy balance algorithms for mapping ET: Current status and future challenges. Trans. ASABE 2007, 5, 1639–1644.

25. Gowda, P.H.; Chávez, J.L.; Colaizzi, P.D.; Evett, S.R.; Howell, T.A.; Tolk, J.A. ET mapping for agricultural water management: Present status and challenges. Irrig. Sci. 2008, 26, 223–237.

26. Elarab, M.; Ticlavilcab, A.; Torres-Ruab, A.; Maslovac, I.; McKeea, M. Estimating chlorophyll with thermal and broadband multispectralhigh resolution imagery from an unmanned aerial system using relevance vector machines for precision agriculture. Int. J. Appl. Earth Obs. Geoinf. 2015, 43, 32–42.

27. Liaghat, S.; Balasundram, S.K. A review: The role of remote sensing in precision agriculture. Am. J. Agric. Biol. Sci. 2010, 5, 50–55.

28. Berni, J.A.J.; Zarco-Tejada, P.J.; Suarez, L.; Fereres, E. Thermal and narrow-band multispectral remote rensing for vegetation monitoring from an unmanned aerial vehicle. IEEE Trans. Geosci. Remote Sens. 2009, 47, 722–738.

29. Berni, J.A.J.; Zarco-Tejada, P.J.; Sepulcre-Cantó, G.; Fereres, E.; Villalobos, F. Mapping canopy conductance and CWSI in olive

orchards using high resolution thermal remote sensing imagery. Remote Sens. Environ. 2009, 113, 2380–2388.

30. Matese, A.; Toscano, P.; Filippo, S.; Gennaro, D.; Genesio, L.; Vaccari, F.P.; Primicerio, J.; Belli, C.; Zaldei, A.; Bianconi, R.; et al. Intercomparison of UAV, aircraft and satellite remote sensing platforms for precision viticulture. Remote Sens. 2015,7, 2971–2990.

31. Ortega-Farías, S.; López-Olivari, R. Validation of a two-layer model to estimate latent heat flux and evapotranspiration over a drip-irrigated olive orchard. Trans. ASABE 2012, 4, 1169–1178.

32. Webb, E.K.; Pearman, G.I.; Leuning, R. Correction of flux measurements for density effects due to heat and water vapour transfer. Q. J. R. Meteorol. Soc. 1980, 106, 85–100.

33. Schotanus, P.; Nieuwstadt, F.T.M.; de Bruin, H.A.R. Temperature measurement with a sonic anemometer and its application to heat and moisture fluxes. Bound. Layer Meteorol. 1983, 26, 81–93.

34. Wilczak, J.M.; Oncley, S.P.; Stage, S.A. Sonic anemometer tilt correction algorithms. Bound. Layer Meteorol. 2001, 99, 127–150.

35. Ortega-Farias, S.; Poblete-Echeverría, C.; Brisson, N. Parameterization of a two layer model for estimating vineyard evapotranspiration using meteorological measurements. Agric. For. Meteorol. 2010, 150, 276–286.

36. Er-Raki, S.; Chehbouni, A.; Hoedjes, J.C.B.; Ezzahar, J.; Duchemin, B.; Jacob, F. Improvement of FAO-56 method for olive orchards through sequential assimilation of thermal infrared-based estimates of ET. Agric. Water Manag. 2008, 95, 309–321.

37. Martínez-Cob, A.; Faci, J.M. Evapotranspiration of an hedge-pruned olive orchard in semiarid area of NE Spain. Agric. Water Manag. 2010, 97, 410–418.

38. Twine, T.E.; Kustas, W.P.; Norman, J.M.; Cook, D.R.; Houser, P.R.; Meyers, T.P.; Prueger, J.H.; Starks, P.J.; Wesely, M.L. Correcting eddy covariance flux underestimates over a grassland. Agric. For. Meteorol. 2000, 103, 279–300.

39. Goward, S.N.; Markham, B.; Dye, D.G.; Dulaney, W.; Yang, J.L. Normalized difference vegetation index measurements from the advanced very high-resolution radiometer. Remote Sens. Environ. 1991, 35, 257–277.

40. Rouse, J.W.; Haas, R.H.; Schell, J.A.; Deering, D.W. Monitoring vegetation systems in the Great Plains with ERTS. Third ERTS Symposium; Freden, S.C., Becker, M.A., Eds.; NASA Goddard Space Flight Center: Greenbelt, MD, USA, 1973; pp. 309–317.

41. Zarco-Tejada, P.J.; Gonzalez-Dugo, V.; Berni, J.A.J. Fluorescence, temperature and narrowband indices acquired from a UAV platform for water stress detection using a micro-hyperspectral imager and a thermal camera. Remote Sens. Environ.2012, 117, 322–337.

42. Corcóles, J.I.; Jose, F.; Ortega, J.F.; Hernández, D.; Moreno, M.A. Estimation of leaf area index in onion (Allium cepa L.) using an unmanned aerial vehicle. Biosyst. Eng. 2013, 115, 31–42.

43. Sanchez, J.M.; Lopez-Urrea, R.; Rubio, E.; Caselles, V. Determining water use of sorghum from two-source energy balance and radiometric temperatures. Hydrol. Earth Syst. Sci. 2011, 15, 3061–3070.

44. Sánchez, J.M.; López-Urreab, R.; Rubioc, E.; González-Piquerasd, J.; Caselles, V. Assessing crop coefficients of sunflower and canola using two-source energy balance and thermal radiometry. Agric. Water Manag. 2014, 137, 23–29.

45. Kustas, W.P.; Alfieri, J.G.; Anderson, M.C.; Colaizzi, P.D.; Prueger, J.H.; Evett, S.R.; Neale, C.; French, A.; Hipps, L.E.; Chávez, J.L.; et al. Evaluating the two-source energy balance model using local thermal and surface flux observations in a strongly advective irrigated agricultural area. Adv. Water Resour. 2012, 50, 120–133.

46. Poblete-Echeverría, C.; Ortega-Farias, S. Parameterization of surface energy balance components over a drip-irrigated Merlot vineyard using reflectance and meteorological data. Irrig. Sci. 2012, 30, 485–497.

47. Long, D.; Singh, V.P. A modified surface energy balance algorithm for land (M-SEBAL) based on a trapezoidal framework.Water Resour. Res. 2012, 48.

48. Shuttleworth, W.J.; Wallace, J.S. Evaporation from sparse crops-an energy combination theory. Q. J. R. Met. Soc. 1985, 111, 839–855.

49. López-Olivari, R.; Ortega-Farías, S.; Morales, L.; Valdés, H. Evaluation of three semi-empirical approaches to estimate the net radiation over a drip-irrigated olive orchard. Chil. J. Agric. Res. 2015, 75, 341–349.

50. Gonzalez-Dugo, V.; Zarco-Tejada, P.; Berni, J.A.J.; Suárez, L.; Goldhamer, D.; Fereres, E. Almond tree canopy temperature reveals intra-crown variability that is water stress-dependent. Agric. For. Meteorol. 2012, 154–155, 156–165.

51. Willmott, C.J. On the validation of models. Phys. Geogr. 1981, 2, 184–194.

52. Mayer, D.G.; Butler, D.G. Statistical validation. Ecol. Model. 1993, 1–2, 21–31.

53. Fernandes-Silva, A.A.; Ferreira, T.C.; Correia, C.M.; Malheiro, A.C.; Villalobos, F.J. Influence of different irrigation regimes on crop yield and water use efficiency of olive. Plant Soil 2010, 333, 35–47.

54. Ezzahar, J.; Chehbouni, A.; Hoedjes, J.C.B.; Er-raki, S.; Chehbouni, A.H.; Bonnefond, J.M.; De Bruin, H.A.R. The use of the scintillation technique for estimating and monitoring water consumption of olive orchards in a semi-arid region. Agric. Water Manag. 2007, 89, 173–184.

55. Er-Raki, S.; Chehbouni, A.; Boulet, G.; Williams, D.G. Using the dual approach of FAO56 for partitioning ET into soil and plant components for olive orchards in a semiarid region. Agric. Water Manag. 2010, 97, 1769–1778.

56. Villalobos, F.J.; Orgaz, F.; Testi, L.; Federes, E. Measurement and modeling of evapotranspiration of olive (Olea europaea L.) orchards. Eur. J. Agron. 2000, 13, 155–163.

57. Testi, L.; Orgaz, F.; Villalobos, F.J. Variations in bulk canopy conductance of an irrigated olive (Olea europaea L.) orchard.Environ. Exp. Bot. 2006, 55, 15–28.

58. Lee, X.; Black, T.A. Atmospheric turbulence within and above a douglas-fir stand. Part II. Eddy fluxes of sensible heat and water vapour. Bound. Layer Meteorol. 1993, 64, 369–389.

59. Wilson, K.; Goldstein, A.; Falge, E.; Aubinet, M.; Baldocchi, D.; Berbigier, P.; Bernhofer, C.; Ceulemans, R.; Dolman, H.; Field, C.; et al. Energy balance closure at FLUXNET sites. Agric. For. Meteorol. 2002, 113, 223–243.

60. Leuning, R.; van Gorsel, E.; Massman, W.J.; Isaac, P.R. Reflections on the surface energy imbalance problem. Agric. For. Meteorol. 2012, 156, 65–74.

61. Williams, D.G.; Cable, W.; Hultine, K.; Hoedjes, J.C.B.; Yepez, E.A.; Simonneaux, V.; Er-Raki, S.; Boulet, G.; de Bruin, H.A.R.; Chehbouni, A.; et al. Evapotranspiration components determined by stable isotope, sap flow and eddy covariance techniques. Agric. For. Meteorol. 2004, 125, 241–258.

62. Testi, L.; Villalobos, F.J.; Orgaz, F. Evapotranspiration of a young irrigated olive orchard in southern Spain. Agric. For. Meteorol. 2004, 121, 1–18.

63. Foken, T. The energy balance closure problem: An overview. Ecol. Appl. 2008, 18, 1351–1367.

64. Chávez, J.; Gowda, P.; Howell, T.; Neale, C.; Copeland, K. Estimating hourly crop ET using a two-source energy balance model and multispectral airborne imagery. Irrig. Sci. 2009, 28, 79–91.

65. Jusoff, K. Evaluation of spatial variability of soil in an oil palm plantation. In Proceedings of the ISCO 2004—13th International Soil Conservation Organization Conference, Brisbane, Australia, 4–8 July 2004.

66. Liu, S.; Lu, L.; Mao, D.; Jia, L. Evaluating parameterizations of aerodynamic resistance to heat transfer using field measurements. Hydrol. Earth Syst. Sci. 2007, 11, 769–783.

67. Ortega-Farias, S.; Aguilar, R.; De la Fuente, D.; Ortega-Salaza, S.; Fuentes, F. Evaluation of a model to estimate net radiation over a drip-irrigated olive orchard using Landsat satellite images. Acta Hortic. (ISHS) 2014, 1057, 309–314.

68. Shaomin, L.; Guang, H.; Li, L.; Defa, M. Estimation of regional evapotranspiration by TM/ETM+ data over heterogeneous surfaces. Am. Soc. Photogramm. Remote Sens. 2007, 73, 1169–1178.

69. González-Dugo, M.P.; González-Piqueras, J.; Campos, I.; Andréu, A.; Balbontín, C.; Calera, A. Evapotranspiration monitoring in a vineyard using satellite-based thermal remote sensing. SPIE Proc. 2012, 8531, 272–288.

70. Ortega-Farias, S.; Aguilar, R.; De la Fuente, D.; Ortega-Salaza, S.; Fuentes, F.; Poblete-Echeverría, C. Estimation of evapotranspiration for a drip-irrigated olive orchard using multispectral satellite images. In Proceedings of the USCID

Fourth International Conference on Irrigation and Drainage, Using 21st Century Technology to Better Manage Irrigation Water Supplies, Phoenix, AZ, USA, 16–19 April 2013.

71. Xia, T.; Kustas, W.P.; Anderson, M.C.; Alfieri, J.G.; Gao, F.; McKee, L.; Prueger, J.H.; Geli, H.; Neale, C.M.U.; Sanchez, L.; et al. Mapping evapotranspiration with high-resolution aircraft imagery over vineyards using one- and two-source modeling schemes. Hydrol. Earth Syst. Sci. 2016, 20, 1523–1545.

72. Homann, H.; Nieto, H.; Jensen, R.; Guzinski, R.; Zarco-Tejada, P.J.; Friborg, T. Estimating evaporation with thermal UAV data and two-source energy balance models. Hydrol. Earth Syst. Sci. 2016, 20, 697–713.

4

A SOLAR ENERGY POWERED AUTONOMOUS WIRELESS ACTUATOR NODE FOR IRRIGATION SYSTEMS

Rafael Lajara, Jorge Alberola and José Pelegrí-Sebastiá

Research Institute for Integrated Management of Coastal Areas, Universitat Politècnica Valencia, EPSG, C.Paranimf, 1, Gandía, Spain

ABSTRACT

The design of a fully autonomous and wireless actuator node ("wEcoValve mote") based on the IEEE 802.15.4 standard is presented. The system allows remote control (open/close) of a 3-lead magnetic latch solenoid, commonly used in drip irrigation systems in applications such as agricultural areas, greenhouses, gardens, *etc.* The very low power consumption of the system in conjunction with the low power consumption of the valve, only when switching positions, allows the system to be solar powered, thus eliminating the need of wires and facilitating its deployment. By using supercapacitors recharged from a specifically designed solar power module, the need to replace batteries is also eliminated and the system is completely autonomous and maintenance free. The "wEcoValve mote" firmware is based on a synchronous protocol that allows a bidirectional communication with a latency optimized for real-time work, with a synchronization time between nodes of 4 s, thus achieving a power consumption average of 2.9 mW.

Keywords: wireless sensor networks; actuator node; autonomous sensor; solar energy; agriculture irrigation systems

1. INTRODUCTION

Localized irrigation systems require the use of latch type solenoids to control automated irrigation zones. Currently, control is achieved by using wired networks with two or three leads to communicate and carry out the actions (activation/deactivation) of the solenoid, such as the Piccolo RTU from Motorola [1]. Some newer systems are based on a 2.4 GHz radio link [2–5] but do not implement the control of the solenoid valve. Most of them, normally, are used for monitoring environmental conditions. There are other systems that take control when implemented, such as Piccolo-XR [1] but they need a battery, which must be replaced from time to time in order to assure the action of the solenoid.

Wireless sensor networks have the peculiarities that, as a rule, do not use large data transfer rates, and furthermore, their nodes are powered by small batteries and/or supercapacitors. Our design consists of a hardware part, where the components were chosen to optimize the use of the available energy. The firmware has also been taken into account for this purpose, as the main goal of a MAC layer protocol for these devices is to save energy. This requires to identify the main mechanisms of energy cost as the detailed in [6], hence the importance of timing in this type of nodes and applications.

Traditional MAC protocols keep the receivers always listening to a channel, since energy efficiency is not priority [7], but this is not practical in low power sensor networks, where energy efficiency is a major concern. Thus the sensor network nodes should power-off their transceivers when they have a chance [8]. These reasons and others such as hardware limitations or the need for reconfiguration make traditional synchronization mechanisms not valid in low power sensor networks, and new approaches claim to be developed specifically, even for each type of application in particular [7–13].

For two sensor nodes to communicate with each other, they must enable their transceivers within the same period of time (synchronously), one in transmission mode and the other in reception mode. The problem to be solved by the MAC protocol is to determine when to enable the transceiver, having into account that there must be synchronization between at least two of the nodes. In addition they must also take into account the time when two nodes want to transmit at a time to avoid collisions, and when a node wants to transmit data but must also receive.

The main energy saving is achieved by disabling the transceiver when not needed. For example, the CC2420 transceiver, used in this node consumes 19.7 mA in transmission, 17.4 mA in reception and 20 μA in power-down mode [14]. Of course, the microprocessor must be waked up from its idle or sleep mode before enabling the transceiver. In addition it would also be desirable for other devices such as sensors and/or actuators be disabled when not needed.

Activating devices only when they are needed, leads to the typical listen/sleep duty cycles of low power sensor networks applications. In active mode, the node will perform tasks as transmitting or sensing, and in sleep mode it will turn off unused devices. For simplicity consider that the processor and the sensors have the same activation period of the transceiver. The lower is the duty cycle, the lower is the average consumption.

2. wECOVALVE HARDWARE DEVELOPMENT

The "wEcoValve mote" block diagram is shown in Figure 1, where there are three sections: (a) the supply circuit formed by the solar panel, a supercapacitor and two highly efficient DC-DC converters to supply a regulated 3.3 V to the microcontroller and the transceiver; (b) the control circuit for the solenoid, which comprises a DC-DC converter, a supercapacitor and a circuit to control the solenoid valve with a 12 V supply; (c) diagram of the microcontroller and the transceiver, implemented by a Flexipanel Pixie module.

The actuator node is powered by a circuit similar to the one shown in [15], but adapted to the needs of the action over the solenoid valve and based on the use of a solar panel that recharges a single 100 F supercapacitor to be fully autonomous without replaceable batteries.

The Epson SG-3030JC oscillator and the 74HC40103 counter have been used to divide the clock frequency down and allow the microcontroller to work in the lowest power mode, as explained later. Current pulses of 700 mA and 12 V peak and 20 ms long open and close the solenoid valve [16]. These pulses are achieved through a 12 V Boost Converter and a second 3.3 mF supercapacitor that stores energy for activation/deactivation of the solenoid. The supercapacitor is charged up to 10.5 V, and this voltage is measured by the microcontroller by means of polling. Once the supercapacitor has accumulated enough energy, the triggering circuit of the solenoid valve activates the discharge of the supercapacitor with a 10 μs voltage pulse. Finally, other circuits were added to measure the current provided by the solar panel and to measure the voltage across the main supercapacitor for testing and debugging purposes.

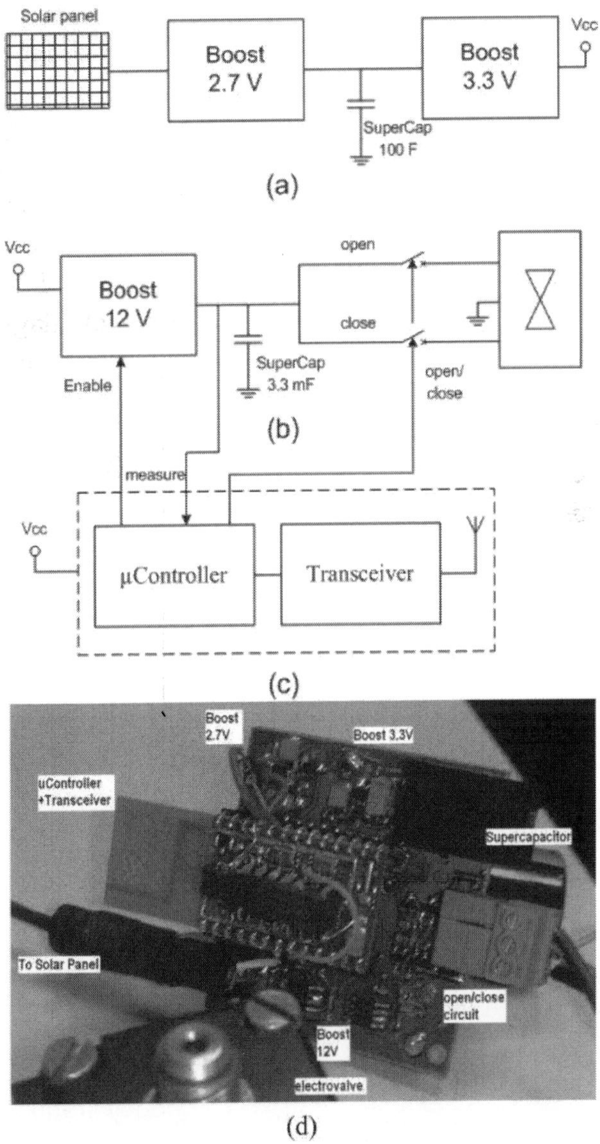

Figure 1. Block diagram of a terminal node. **(a)** Power circuit formed by the solar panel, a supercapacitor and two highly efficient DC-DC converters; **(b)** Control circuit of the electrovalve formed by a DC-DC converter, a supercapacitor and a discharge circuit to control the electrovalve; **(c)** Microcontroller and tranceiver block diagram with a Flexipanel Pixie module; **(d)** A photo of the wEcoValve laboratory prototype.

3. SYNCHRONIZATION PROTOCOL DEVELOPMENT

In [7] a classification of different types of protocols was given, highlighting the slotted (TDMA, 802.15.4, SMAC, DSMAC, TMAC, FRAME), the sampling (ALOHA, BMAC, WiseMAC, CSMA; (for wireless networks using CSMA/CA)) and the others (STEM (using two transceivers, one for data transfer with a certain listen/sleep duty cycle, and the other one that is permanently activated to wake-up the system) and hybrids (e.g., ZMAC, SCP-MAC)).

Slotted protocols keep nodes permanently synchronized, so they are particularly well suited for applications where there is a periodic exchange of information. Also they know exactly when to be activated, allowing them to shorten the active periods. Moreover, sampling protocols will be more convenient in terms of energy in applications with sporadic transmissions, because there is no penalty to maintain the synchronization.

3.1. Slotted Protocols

Slotted protocols require that all nodes work with a common clock signal, which will serve to maintain synchronization. In the end, at the lowest level, all of these protocols use the main crystal of the system to maintain the synchronization. If these clocks were ideal, after initialization, the nodes would always be perfectly synchronized, however frequency variations occur between clocks from one node to another due to various causes, such as temperature changes, aging or frequency stability [13].

For example, the clock used in the Microchip PICDEM.Z evaluation board, the HC-49US from ECS International Inc., has a frequency stability of ±15 ppm to ±100 ppm at 25 °C [17]. This means that at worst, within 1 minute the clocks of two nodes have been lagged 12 ms. To maintain synchronicity one can send special frames (beacons) broadcasted periodically or use the preamble of the messages to send information about the synchronization. In both cases, drift margins have to be taken into account to avoid possible desynchronizations.

MAC protocols focus on activating and deactivating the transceiver, and this requires the microprocessor to maintain the synchronization. This entails a problem if you want to set the microprocessor to a low-power mode: the microprocessor maintains the synchronism, but if all of the devices (including timers) were turned off, the synchronization would get lost.

The PIC18F4620 microcontroller used has a clock frequency of 4 MHz and 3 V draws 1.3 mA in active mode, 430 μA in PRI_IDLE mode (CPU and peripherals switched off) and 0.1 μA deep sleep mode [17]. For a duty cycle of 1% and a 21 mA current consumption in listening mode (microcontroller plus transceiver active), and a current consumption of 450 μA in sleep mode (PRI_IDLE mode microcontroller and transceiver powered down), the overall average consumption is 655.5 μA.

However the use of microcontroller's deep sleep mode (0.1 μA) can achieve an average consumption of 210.1 μA. For that, the aim is to change the microcontroller from one mode to another during its sleep period, thus reducing the consumption by two thirds.

3.2. The Proposal

Slotted protocols with a low duty cycle will approach the average consumption to the sleep mode consumption. Therefore, reducing this consumption has a major impact on the average consumption. The aim is to optimize the power consumption for a given protocol by reducing the current drawn in sleep mode. To do this, one needs to bring the microprocessor to the lowest power mode and stay there as much time as possible. For that, all the peripherals and even the processor must be powered down, all but some system to cause wakeup again.

The first approach would be to use some internal oscillator to periodically wake the processor up. For example, the PIC18F4620 microcontroller incorporates an internal 32 kHz oscillator [18] connected to the watchdog timer, which can be used for this task. But this presents two downsides: 32 kHz is too high a frequency for low duty cycle applications of low power sensor networks, so the processor probably wakes up more often than necessary. This can be solved by using the prescaler and postscaler counters of the watchdog circuit. The second and main problem is that this oscillator is an RC type, which means that it has worse stability than those based on quartz crystals. The microcontroller under study can vary its frequency from 26.562 kHz to 35.938 kHz, which makes a count that may be 1 s long differs up to 0.3 s at the worst case. This problem makes to discard the use of RC oscillators. It is also the reason why so far the quartz crystal and a counter have been used as the main oscillators and as the basis of synchronism.

Instead of using the internal RC oscillator, a simple circuit consisting of a low frequency crystal oscillator and a frequency divider has been added, see Figure 2. The output of this circuit is connected to an input that triggers an interrupt, which wakes up the processor. This solution allows turning all the peripherals of the processor fully off, so that the consumption in sleep mode is considerably reduced.

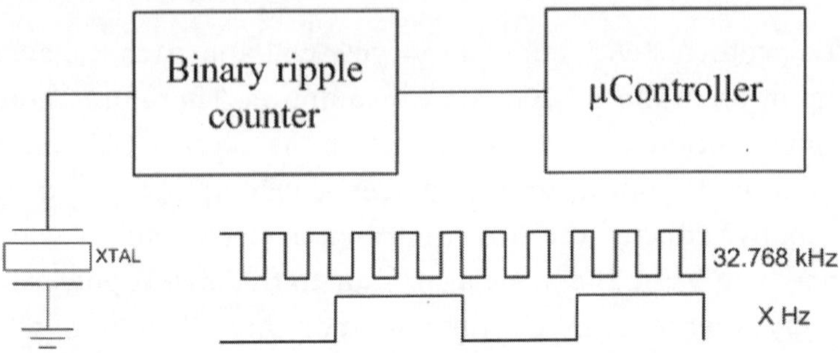

Figure 2. Introduction to a binary ripple counter that permits the microcontroller to enter sleep mode.

However, when using an external oscillator of 32.768 kHz, consumptions and tolerances can be widely improved. The oscillator that we take as an example is the 90SMX from IQD [19], which has a maximum frequency tolerance of ±100 ppm (typically ±20 ppm). The worst case is when two nodes are desynchronized the same time as the preceding example, where there was a 4 MHz clock with a 100 ppm tolerance, *i.e.*, 12 ms over 1 minute.

This accuracy is more than enough for typical applications in low power sensor networks such as drip irrigation. In fact sometimes the processor's wakeup time may be more relevant than the imprecision of the clocking. With this solution, there is sufficient accuracy for the time measurement. Concerning the consumption, the crystal has its maximum value of 1 μW. On the other hand the PIC18F4620 microcontroller can be brought to its lowest power consumption mode, so-called sleep (OSCON = 0), whose consumption is 0.3 μW (0.1 μA at 3 V).

By adding a frequency divider like the 74HC40103 for the oscillator, a maximum current of 8 μA at 6 V must be added. The use of this divider allows the microcontroller to enter the sleep mode most of the time, saving energy (see Figure 2). Therefore there is a power consumption of 1 μW (oscillator) + 0.3 μW (microcontroller in sleep) + 12 μW (frequency divider at 3 V). That is a total of 13.3 μW.

A typical synchronization scheme would save the low frequency external clock and the frequency divider, but the processor could not enter the deepest sleep mode. The previous microcontroller would have to use an internal counter to be active, therefore it would have to enter the PRI_IDLE mode. This entails a consumption of 430 μA at 3 V and 4 MHz for the CPU and 2.6 μA for the internal counter of the microcontroller. It would be 1.3 mW in total.

The power consumption of the transceiver in sleep mode and the rest of the device should be added in both measures, but the consumption is reduced by 100 times only considering the components that affect the synchronization.

4. "wECOVALVE" FIRMWARE/SOFTWARE DEVELOPMENT

The solution proposed in Section 3.2 has been implemented together with the SMAC protocol [20]. This protocol allows two or more nodes to synchronize and communicate with each other during their active period. The synchronization protocol is accomplished with regular broadcast transmission of SYNC frames from nodes to nodes; the nodes update their timers to keep synchronized by taking the time from these frames. A scheme based on containment through frames CTS/RTS is used to avoid collisions. RTS frames are a request to send data and CTS is the confirmation. Data frames are sent after that if needed. When there is no data to broadcast, the node enters sleep state. A scheme of these frames can be seen in Figure 3.

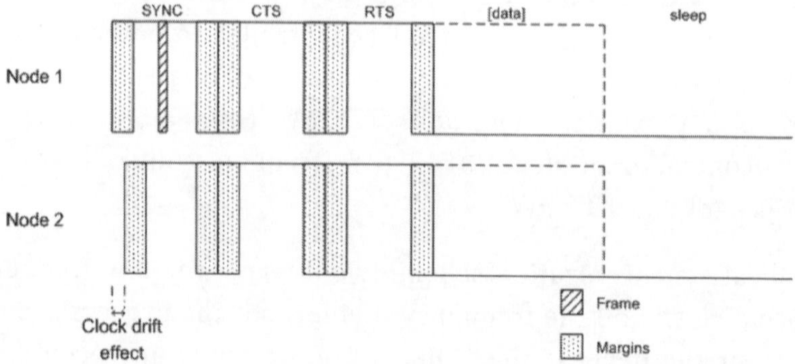

Figure 3. Scheme listen/sleep from two nodes. The second node starts its listen period after the listen period of the first node. When node 1 broadcasts the SYNC frame, both nodes get resynchronized. The nodes reserve margins (dotted area) in each sub-period (SYNC, CTS and RTS) of listen mode in order to avoid the loss of a frame when other nodes are still not listening. The frames are sent passed a random time, after the beginning of each sub-period, respecting the safe margins. When a node receives a frame in a sub-period, it must await for the next listen period.

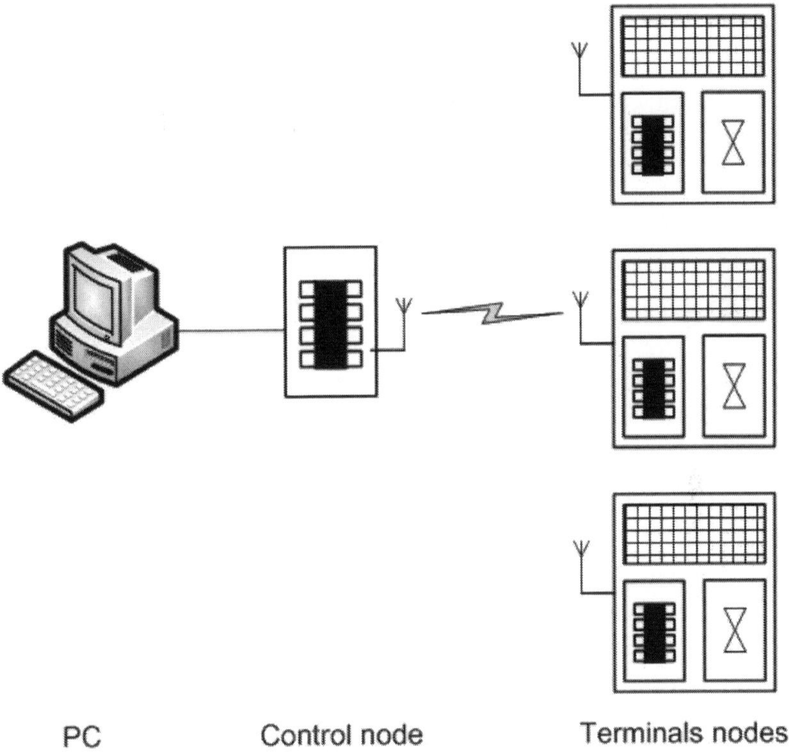

Figure. 4. Scheme of communications between the "control nodes" and "terminals nodes".

Some of the nodes that form the network are called "control nodes", and transmit the open and close commands to remotely control the solenoid valves. Other nodes are called "terminal nodes", and are the ones that receive the commands for actuating over the solenoid valves. The latter shall be responsible for monitoring the valve states (cf. Figure 4). Furthermore, to assist in debugging the application, "terminal nodes" broadcast the current value supplied by the solar panel and the voltage value across the supercapacitors. This information also aids to know the status of each node at any moment. Thus it could be used to know whether a node might fail for not being able to open/close the solenoid, or might not become active again due to the lack of energy. This information is received by the "control node", and passed through a serial communication to an application running on a PC. This application allows the user to broadcast opening or closing commands from a "control node" to each valve.

5. RESULTS AND DISCUSSION

The remotely controlled "wEcoValve" has been implemented by using the SMAC protocol. In the implementation a 4 seconds period of resynchronization has been chosen, see Figure 5. Every 4 seconds a node broadcasts a synchronization frame to be listened by all its neighbors. Since there is a random delay before a node broadcasts a SYNC frame, the node with the shortest delay updates the time of the rest. Thus, the node which transmits the SYNC frame remains uncertain, and it is randomly selected.

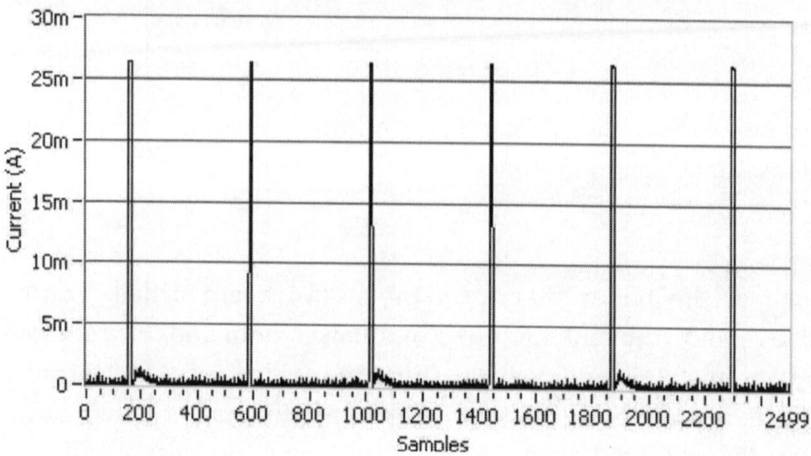

Figure 5. Current consumption of a terminal node. The peaks of 27 mA are produced every 4 seconds due to the synchronization between nodes. The average consumption is 836.76 μA.

The resynchronization time can be increased, whereas the clock's drift, not evident. In the same manner this time also can be reduced to improve the latency. A monitoring application has also been implemented on the PC, which can send data via RS232 to a "control node" in order to open and close the "terminal nodes". This can be scheduled.

Experimental measurement results of the current generated by the solar panel and the voltage by the supercapacitor are shown in Figures 6 and 7. In Figure 6 the supply supercapacitor discharge curve of the wEcoValve mote has a lifetime of approximately 26 hours without solar energy support, with a synchronization period of 4 seconds. A lifetime of 50 hours can be reached with a synchronization period of 8 seconds. Figure 7 presents the supercapacitor discharge curve (green) for two consecutive opening and closing actions of the solenoid (in blue).

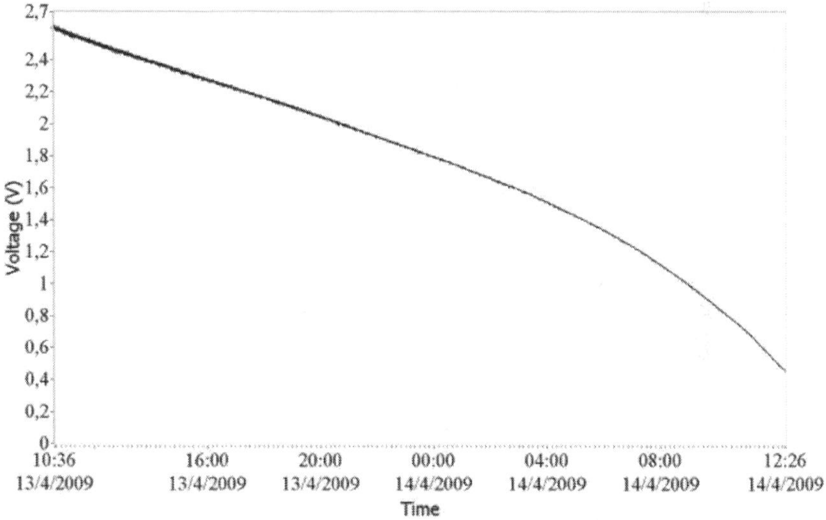

Figure 6. Supply supercapacitor discharge curve of the wEcoValve mote. It has a lifetime of approximately 26 hours without solar energy support. The synchronization period is set to 4 seconds. A lifetime of 50 hours can be reached with a synchronization period of 8 seconds.

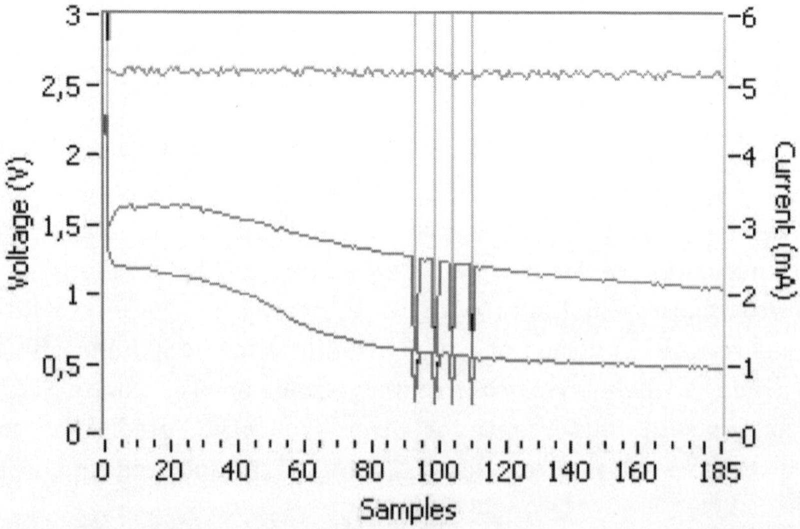

Figure 7. Supercapacitor discharge (green) for two consecutive opening and closing actions of the solenoid (in blue).

On the other hand the hardware developed (Figure 1(b)) to open/close the solenoid has been optimized so that the charging and discharging of the supercapacitor of 3.3 mF only occurs when requested and by using as little energy as possible. This can be seen in Figure 8, where a short pulse (microseconds) initiates the charge of the boost's capacitors up to the opening/closing threshold.

Once the threshold is reached (about 11.4 V), a 15 ms pulse is long enough to open the valve, and this action drops about 3 V (Figure 8(a) open). Figure 8(b) is closed; a 20 ms pulse is long enough to close the valve, with a voltage drop of about 5 V. Figure 9 shows the current necessary for the activation and deactivation of the solenoid.

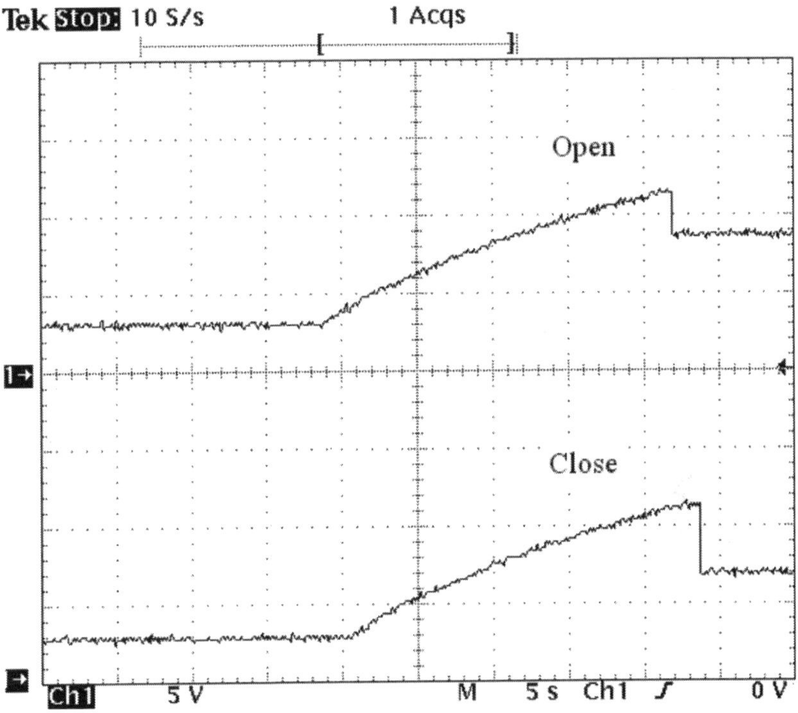

Figure 8. (a) Open. A short pulse (microseconds) initiates the charge of the boost's capacitors up to the opening/closing threshold. Once the threshold is reached (about 11.4 V), a 15 ms pulse is long enough to open the valve, and this action drops about 3 V. **(b)** Close. A 20 ms pulse is long enough to close the valve, with a voltage drop of about 5 V.

Neither the opening nor the closing actions fully discharge the boost's capacitor (3.3 mF), leaving some residual charge which might vary from 7 V to 9 V, thus allowing for quicker consecutive loads with delays of only 3 s in the worst case, unlike a complete charging time that takes 23.1 seconds for both opening and closing the solenoid valve. The voltage threshold that allows fully triggering the valve is approximately 11.4 V and the voltage drop caused by the switching normally is 2.8 V when opening and 4.6 V when closing.

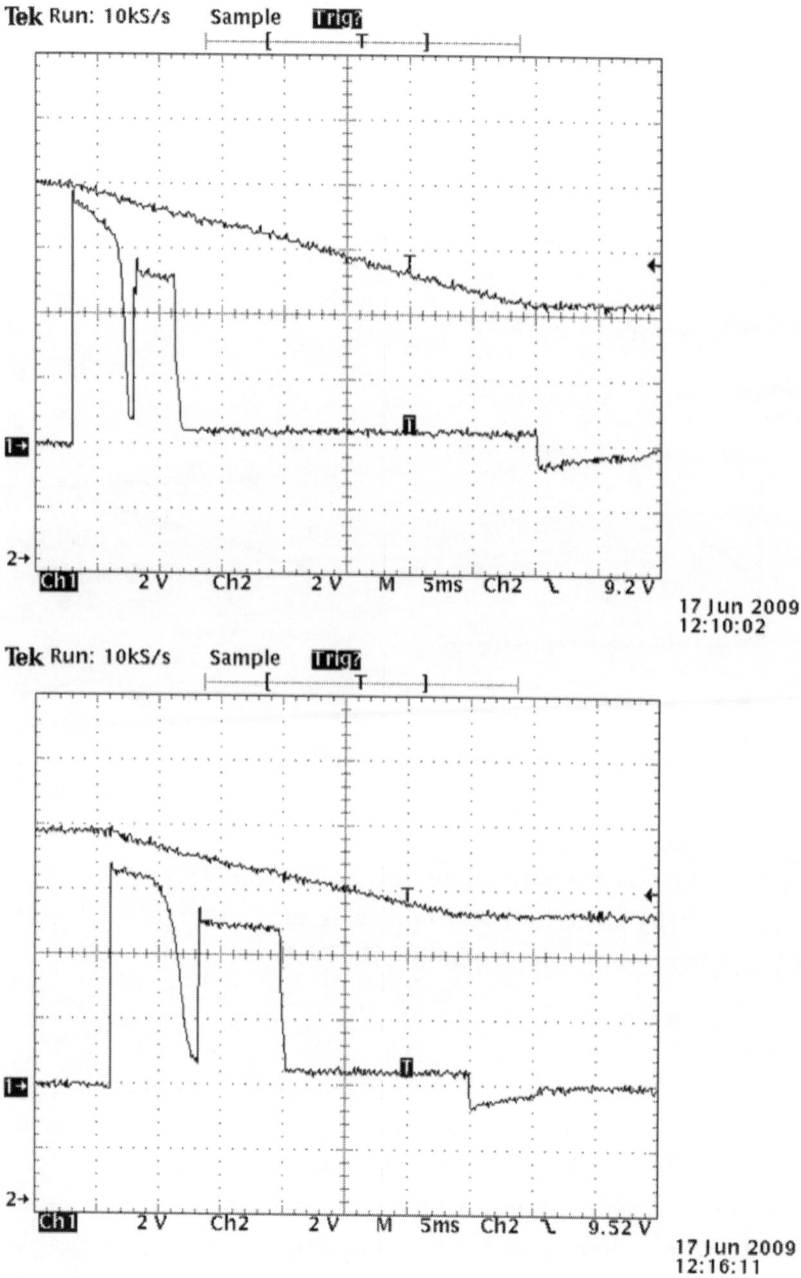

Figure 9. Ch1 is the current that triggers the solenoid and ch2 is the voltage of the capacitor of 3.3 mF.

6. CONCLUSIONS

A fully autonomous and remote solenoid valve node (wEcoValve) that can be used for irrigation of gardens, indoor plants, greenhouses or field crops has been implemented and tested. It is in general suitable for any system that requires a solenoid type latch valve.

Its key advantage is that it does not need any cabling deployment to power the node and the electrovalve, because it is wirelessly controlled and solar powered. This was accomplished with an ultra low power consumption hardware/firmware design that allowed supercapacitors (100 F) to provide the system's power supply for more than a day without direct sunlight over the solar panel, depending on the synchronization period and the duty cycle of use in the end application. The use of supercaps with almost no degradation over time eliminates the need for batteries replacement.

A very precise synchronization method (4 seconds for resynchronization) has been designed by making use of a low frequency crystal and a frequency divider, which let us to shut down most of peripherals, thus reducing considerably the power consumption. It can be used in several synchronization protocols to reduce messaging traffic or improve accuracy. It could be used to open or close electrovalves that control the water flow through plants. Each solenoid could be connected to a node that could be wirelessly and independently controlled.

As future work, we have started to test the board in a real scenario (orange fields in south Valencia) and collect data. Later the system could be adapted for measuring the water flow and to keep track of the consumed water, plus transmitting the data out to a central station.

ACKNOWLEDGMENTS

This work was supported by the I + D + i program of the Generalitat Valenciana, R&D Project GV05/043, and the Vicerecorate of Investigation, Development and Innovation of Universidad Politecnica de Valencia PAID-06-06-002-61 and PAID-10-11.

REFERENCES

1. Irrinet for a Freener World. Irrinet System Description. Product Brief; Schaumburg, IL, USA: Motorola, 2009. Available online: www.mottech.com (accessed on 15 March 2010).

2. Lopez, J.A.; Soto, F.; Sanchez, P.; Iborra, A.; Suardiaz, J.; Vera, J.A. Development of a sensor node for precision horticulture. Sensors 2009, 9, 3240–3255.

3. eKo Outdoor Wireless System for Environmental Monitoring; Crossbow Technology Inc: Milpitas, CA, USA, 2009. Available online: http://www.xbow.com/Products/Product_pdf_files/Wireless_pdf/eko_sensors_Datasheet (accessed on 7 February 2010).

4. Kim, Y; Evans, R.G.; Iversen, W.M. Remote sensing and control of an irrigation system using a distributed wireless sensor network. IEEE Trans. Instrum. Meas 2008, 57, 1379–1387.

5. Chavez, J.L.; Pierce, F.J.; Eliott, T.V. A remote irrigation monitoring and control system for continuous move systems. Part A: Description and development. Precis. Agric 2010, 11, 1–10.

6. Demirkol, I.; Ersoy, C.; Alagoz, F. MAC protocols for wireless sensor networks: A survey. IEEE Commun. Mag 2006,44, 115–121.

7. Kuorilehto, M.; Kohvakka, M.; Suhonen, J.; Hämäläinen, P.; Hännikäinen, M.; Hamalainen, T.D. Ultra-Low Energy Wireless Sensor Networks in Practice; Wiley: Hoboken, NJ, USA, 2007.

8. Sichitiu, M.L.; Veerarittiphan, C. Simple, accurate time synchronization for wireless sensor networks. Wirel. Commun. Netw 2003, 2, 1266–1273.

9. Elson, J.; Römer, K. Wireless sensor networks: A new regime for time synchronization. Sigcomm. Comput. Commun. Rev 2003, 33, 149–154.

10. Faizulkhakov, Y.R. Time synchronization methods for wireless sensor networks: A survey. Program. Comput. Soft 2007,33, 214–226.

11. Sommer, P.; Wattenhofer, R. Gradient clock synchronization in wireless sensor networks. Proceedings of ACM/IEEE International Conference on Information Processing in Sensor Networks (IPSN), San Francisco, CA, USA, 13–16 April 2009.

12. Sundararaman, B.; Buy, U.; Kshemkalyani, A.D. Clock synchronization for wireless sensor networks: A survey. Ad Hoc Netw 2005, 3, 281–323.

13. CC2420, 2.4 GHz IEEE 802.15.1/ZigBee-Ready RF Transceiver Data Sheet; Texas Instrument: Dallas, TX, USA, 2009.

14. Alberola, J.; Pelegri, J.; Lajara, R.; Perez, J.J. Solar inexhaustible power source for wireless sensor node. Proceedings of IEEE International Conference on Instrumentation and Measurement Technology, IMTC 2008, Victoria, Canada, 12–15 May 2008; pp. 657–662.

15. Pelegri, J.; Ramírez, D. Circuit latches solenoid at a distance. EDN Mag 1999, 10, 154.

16. HC-49US Quartz Crystal Datasheet; ECS International Inc: Olathe, KS, USA. Available online: http://www.ecsxtal.com/store/pdf/hc_49us (accessed on 17 October 2010).

17. Enahnced Flash Microcontroller with 10-Bit A/D and nanoWatt Technology Data Sheet, PIC18F4620;. Available online: http://www.chipcatalog.com/Microchip/PIC18F4620.htm (accessed on 16 November 2010.).

18. 90SMX & 91 SMX Crystals Data sheet from Datasheet; IQD Frequency Products: Crewkerne, UK, 2009.

19. Lajara, R.; Pelegri-Sebastia, J.; Perez-Solano, J.J. Power consumption analysis of operating systems for wireless sensor networks. Sensors 2010, 10, 5809–5826.

20. Wei, Y.; Heidemann, J.; Estrin, D. An energy-efficient MAC protocol for wireless sensor networks. Proceedings of Twenty-First Annual Joint Conference of the IEEE Computer and Communications Societies (INFOCOM 2002), New York, NY, USA, June 2002; pp. 1567–1576.

5

EFFECTS OF DRIP IRRIGATION CIRCUIT DESIGN AND LATERAL LINE LENGTHS: I—ON PRESSURE AND FRICTION LOSS

Mohamed Tayel[1*], David Lightfoot[2], Hani Mansour[1]

[1]WRFI Department, National Research Center (NRC), Giza, Egypt

[2]Southern Illinois University Carbondale, Carbondale, USA

ABSTRACT

Laboratory tests were conducted at the Irrigation Devices and Equipment's Test Laboratory, Agricultural Engineering Research Institute, Agriculture Research Center, Giza, Egypt. The experimental design of laboratory experiments was split in randomized complete block design with three replicates. Laboratory tests carried out on three irrigation lateral lines of 40, 60, 80 m under the following three drip irrigation circuit (DIC) designs; 1) one manifold for lateral lines or closed circuits with one manifold of drip irrigation system (CM_1DIS); 2) closed circuits with two manifolds for lateral lines (CM_2DIS), and 3) traditional drip irrigation system (TDIS) as a control. The aims of the work were to study the effect of drip irrigation circuits (DIC) and lateral lines lengths (LLL; where): (LLL_1 = 40 m, LLL_2 = 60 m, and LLL_3 = 80 m) on pressure head (PH) and friction loss (FL). Regarding to LLL and according to PH values, DIC designs could be ranked in the following ascending order: TDIS < CM1DIS < CM2DIS. The differences in PH among DIC designs were significant at the 1% level. The depressive effects of LLL on PH could be ranked in the following ascending order: $LLL_1 < LLL_2 \leq LLL_3$. Differences in PH among LLL treatments were significant at the 1% level except that between LLL_2 and LLL_3. The effects of interactions among: DIC × LLL on PH were significant at the 1% level with some exceptions. The highest value of PH (9.5 m) and the lowest one (6.05 m) were achieved in the interactions of $CM_2DIS \times LLL_1$ and $TDIS \times LLL_3$, respectively. The shapes of the energy gradient lines were affected by DIC and LLL treatments used through their effect on ΔH/H ratio. However, they followed similar trends. According to the FL values, DIC and LLL treatments could be ranked in tha following descending orders TDIS > CM1DIS > CM2DIS and $LLL_1 > LLL_2 > LLL_3$. The differences in FL among DIC and LLL were significant and the effects of interactions among DIC × LLL on FL were significant at the 1% level. The maximum and minimum values of FL were obtained in the interactions: $TDIS \times LLL_3$ and $CM_2DIS \times LLL_1$, respectively. Therefore, the CM_2DIS system is recommended for use where technically feasible.

Keywords: Drip; Irrigation; Circuit; Laterals; Pressure; Friction

1. INTRODUCTION

Differences in emitter geometry may be caused by variation in injection pressure and heat instability during their manufacture, as well as by a heterogeneous mixture of materials used for their production [1]. One some of the factors affecting drip irrigation design was inlet pressure. It was one of the most important factors in drip irrigation design. If the inlet pressure head becomes greater than the required pressure head; it may cause water back-flow and if the inlet pressure head becomes lower than the total required pressure head, it may create negative pressure at the lateral which will affect distribution uniformity. Consequently, to avoid both problems, the inlet pressure head must be determined precisely to balance the energy gain due to inlet flow and the total required pressure head within the lateral line [2]. [3,4] attempted a mathematical approach to calculate the inlet pressure head. In any irrigation system, energy required for system operation depends on the required head and the system discharge. [5] used the relationship:

$$q_e = kH^x \tag{1}$$

where: q_e is the emitter flow rate (L^3t^{-1}); k is the emitter constant; x is the pressure head exponent; and H is pressure head (L).

[6] indicated that the relation between the flow rate and the pressure head is nonlinear in the transition and the turbulent flow types. Also he proposed a method to incorporate pipe components into the hydraulic network analysis by adding their contribution to the nodal equations instead of treating them as separate items. [7] used the Darcy-Wiesbach equation to evaluate the friction losses in a plastic pipe. He expressed the friction loss in the pipe as follows:

$$h_{loss} = 8 f_s Ql / \pi g D^2 \tag{2}$$

where: h_{loss} = Head loss (m), f_s = the coefficient of friction (m/100 m); Q = the flow moving through the pipe ($l \cdot h^{-1}$); l = the pipe length (m); g = the gravitational acceleration (m/sec^{-2}); and D = the pipe inside diameter (mm).

[2] used the Darcy-Wiesbach equation and calculated the value of f_s. Based on the work of [7,8] they used their equation to calculate the friction coefficient based on the flow type being laminar, transient or turbulent. The local head loss is mainly due to friction losses in PE tubes and changes in water temperature in the lateral. Friction loss due to the velocity of water can be determined using Darcy-Weisbach equation. Although a single emitter generally produces a small local loss but due to the high number of emitters installed along a lateral, the total amount of local losses can become a significant fraction of the total energy loss [9]. [10] found that the

head loss in elbows, tees, and valves can significantly affect the pressure in an irrigation network. [11] developed a computer model to optimize the irrigation system design for small areas in South Dakota, USA. The model considers crop type, soil type, irrigation interval, system layout, and pressure requirements of the emitter. Some of the parameters needed for the system design were calculated using the generalized equation for predicting parameters, such as the wetting diameter, the shortest irrigation interval, etc.

The manuscript aims to study the effect of drip irrigation circuits (DIC) used: 1) Closed irrigation circuit with one manifold for lateral lines (CM_1DIS) 2) Closed irrigation circuit with two manifolds for lateral lines (CM_2DIS), 3) traditional drip irrigation system (TDIS) as a control and lateral lines lengths (LLL): (LLL_1 = 40 m, LLL_2 = 60 m, LLL_3 = 80 m) on: pressure head and friction loss.

2. MATERIAL AND METHODS

The laboratory tests were conducted at Irrigation Devices and Equipments Tests Laboratory, Agricultural Engineering Research Institute, Agriculture Research Center, Giza, Egypt. The experimental design of laboratory experiments was split in randomized complete block design with three replicates. Laboratory tests carried out on three irrigation lateral lines 40, 60, 80 m under the following three drip irrigation circuits (DIC) of: 1) one manifold for lateral lines or closed circuits with one manifold of drip irrigation system (CM_1DIS); 2) closed circuits with two manifolds for lateral lines (CM_2DIS), and 3) traditional drip irrigation system (TDIS) as a control, Figures 1-4 showed the directions of flow inside manifold and lateral tubes in the different DIC tested. Details of the pressure and water supply control have been described by [12]. Tests had been carried out in order to resolve the problem of lack of pressure head at the end of lateral lines in the TDIS.

Irrigation networks shown in Figures 1-3 included the following components: 1) Control head: It was located at the water source supply. It consists of a centrifugal pump 3"/3", driven by electric

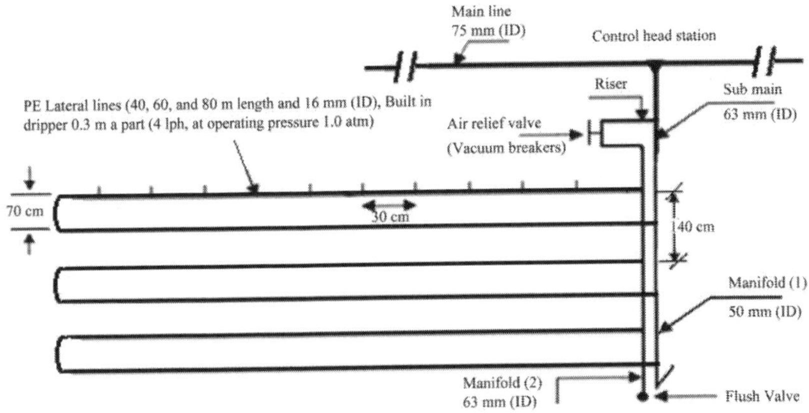

Figure 1. Layout of drippers in a closed circuit design with two manifolds (CM_2DIS) for the lateral lines.

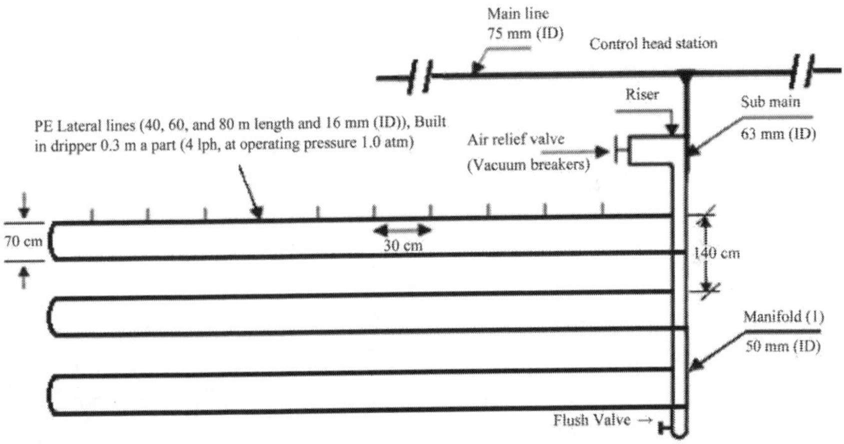

Figure 2. Layout of drippers in a closed circuit design with one manifold (CM_1DIS) for the lateral lines.

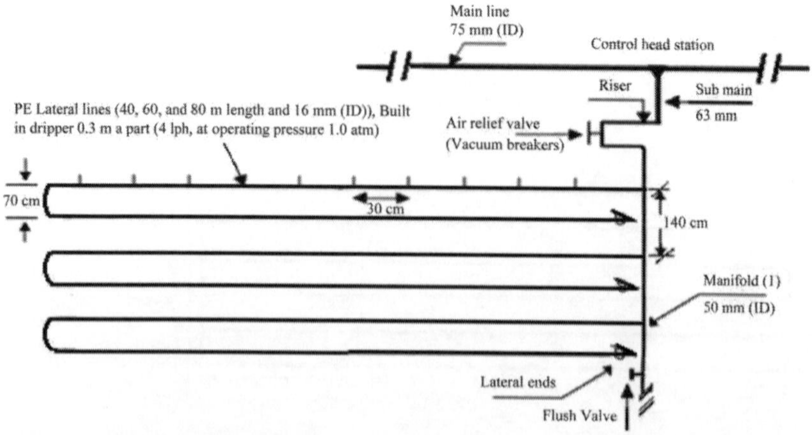

Figure 3. Layout of traditional drip irrigation system (TDIS).

engine (pump discharge of 80 m^3/h and 40 m lift), sand media filter 48" (two tanks), screen filter 2" (120 mesh), back flow prevention device, pressure regulator, pressure gauges, flow-meter, control valves and chemical injection port. 2) Main line: PVC pipes of 75 mm in (ID) to convey the water from the source to the main control points in the field. 3) Submain lines: PVC pipes of 75 mm in (ID) were connected to with the main line through a control unit consists of a 2" ball valve and pressure gauges. 4) Manifold lines: PVC pipes of 50 mm in (ID) were connected to the sub main line through control valves 1.5". 5) Lateral lines: PE tubes of 16 mm in (ID) were connected to the manifolds through beginnings stalled on manifolds lines. 6) Emitters: These emitters (GR) are built in PE tubes 16 mm in (ID), emitter discharge 4 $l \cdot h^{-1}$ at 1 atm. nominal operating pressure and 30 cm spacing in-between. The components of closed circuits of the drip system include, supply lines, control valves, supply and return manifolds, drip lateral lines, emitters, check valves and air relief valves/vacuum breakers [13].

The flow rate through the pipe depends on pipe surface roughness and air layer resistance. The change of hydraulic friction coefficient values, depending on variations in Re number values. Hydraulic losses at plastic pipes might be calculated as losses at hydraulically smooth pipes, multiplied by correction coefficients that assess losses at pipe joints and air resistance.

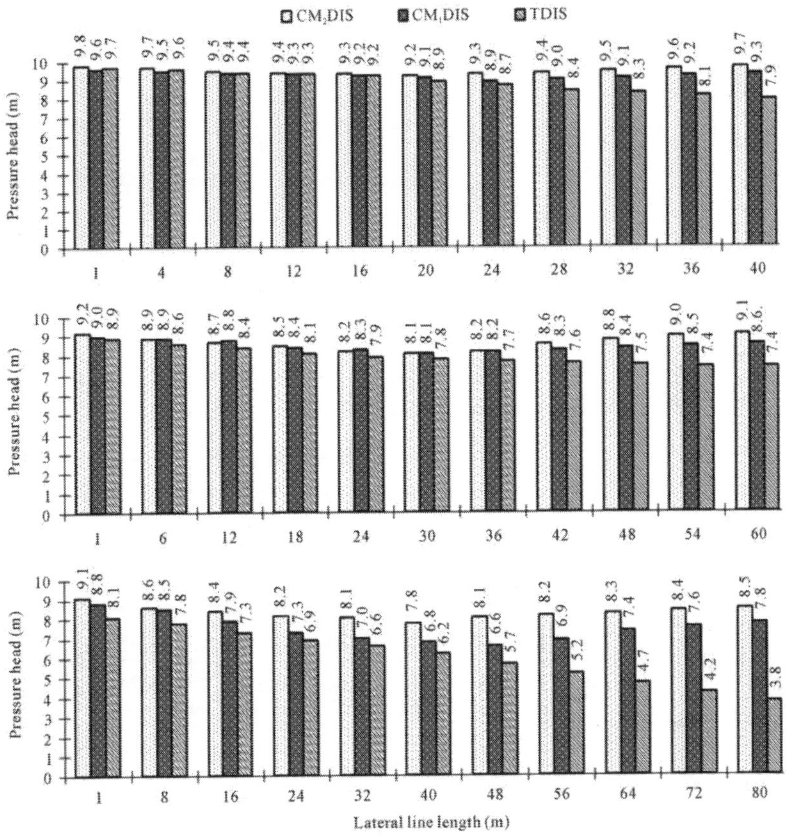

Figure 4. Effect of different irrigation circuits designs on pressure head along different lateral line lengths under (operating pressure = 1.0 atm and slope = 0%).

The energy loss (or head loss) in pipes due to water flow is inversely proportional to the pipe's length.

$$J = \frac{\Delta H}{L} \times 100 \tag{3}$$

where J = The head loss coefficient in a pipe is usually (%) or m/100 mΔH = change in water head (m), and L = length of tube (m).

Coefficient of friction loss was given by [14,15]. The head loss due to friction is calculated by Hazen-Williams equation:

$$\Delta H = \frac{JL}{100} = 1.21 \times 10^{10} \left(\frac{Q}{C}\right)^{1.852} LD^{-4.87} \tag{4}$$

whereΔH = Head loss due to friction (m), J = coefficient of head loss (m/100 m) or %, Q = flow rate is (m^3/h)L = pipe length (m), D = (inner diameter) ID Ø of a pipe (mm), and C = (Hazen-Williams coefficient) smoothness (the roughness) of the internal pipe, (the range for a commercial pipe is 80 - 150). For polyethelene tubes when ID < 40 mm C = 150 [14,15]. [16] stated that head loss due to friction was calculated using the following Darcy-Weisbach equation:

$$h = f(L/D) \times (V^2/2\,g) \tag{5}$$

where h = head loss, m; f = friction factor ; L = length of pipe, m; D = ID Ø of pipe work, m; v = velocity of fluid, m/s; g = acceleration due to gravity, m/s^{-2}.

Friction factor can be expressed as:

$f = 64/R_e$ (For $R_e \leq 2000$)(6)

$f = 0.32 \times R_e^{-0.25}$ (For $R_e \geq 2000$)(7)

where R_e = Reynolds' number, which can be expressed as:

$R_e = vD/\mu$ (8)

Where v = fluid velocity, m/sec; D = ID Ø of lateral, m; and μ = kinematic viscosity of water = 1 × 10^{-6} m^2/sec, at 20°C.

Velocity v (m/s) can be expressed as:

$$v = Q/A \tag{9}$$

where, Q = lateral flow rate ($m^3 \cdot sec^{-1}$) (average flow rate per emitter × number of emitters), and A = cross sectional area of lateral (m^2).

MSTATC program (Michigan State University, East Lansing, MI, USA) was used to carry out statistical analysis. Treatments mean were compared using the technique of analysis of variance (ANOVA) and the least significant difference (L.S.D) between systems at 1% [17].

3. RESULTS AND DISCUSSION

Table 1 and Figures 4 and 5 showed the effect of the DIC used; closed DIC having two and/or one manifolds (CM_2DIS; CM_1DIS), TDIS and LLL (LLL_1 = 40 m, LLL_2 = 60 m; LLL_3 = 80 m) on the parameter under investigation. It can be noticed that PH decreased along the LLL up to 50% - 60% of its length then, it increased again to reach nearly its inlet head pressure in both CM_2DIS and CM_1DIS. On the other hand, it decreased continuously with distance from lateral line inlet in TDIS. This may be due to the existence of two inlets in both CM_2DIS and CM_1DIS which cut down the LLL by about 50% - 60%. According to the Hazen-Williams equation; there is a direct relation between LLL and friction loss. Differences in PH between CM_2DIS and CM_1DIS may be explained on the basis that lateral lines are supplied with water from two manifolds and one manifold, respectively. In other words, the inlet pressure was higher in CM_2DIS relative to CM_1DIS, due to doubling the cross sectional area of the manifolds and that they are connected in parallel in CM_2DIS. Whereas in CM_1DIS, manifold is connected in series and both manifold length (L_m) and resistance increased (Figures 1 and 2).

Regardless of LLL, and according the PH values, DIC used could be ranked in the following ascending order: TDIS < CM_1DIS < CM_2DIS. Difference in PH between any two DIC values was significantly at the 1% level.

Concerning the depressive effect of LLL on PH, LLLs could be ranked in the following ascending order: $LLL_1 < LLL_2 = LLL_3$. Differences in PH between LLL_1 from one side and both LLL_2 and $LLL_3$from the other one were significant at the 1% level. This is due to the direct relation between friction and both lateral line discharge and its length.

Table 1. Effect of drip irrigation circuits (DIC) on pressure head and friction loss (operating pressure = 1 atm and slope = 0%).

ICD	LLL	Pressure head (m)	Friction loss (m)
	40	9.50 a	0.50 i
CM2DIS	60	8.70 dc	1.30 f
	80	8.30 fe	1.70 d
	40	9.23 b	0.80 h
CM1DIS	60	8.33e	1.70 e
	80	7.50 h	2.50 b
	40	8.86 c	1.14 g
TDIS	60	7.99 g	2.21 c
	80	6.05 i	4.00 a
$LSD_{0.01}$ X		0.05	0.02
	CM_2DIS	8.83 a	1.17 c
Means	CM_1DIS	8.35 b	1.67 b
	TDIS	7.63 c	2.45 a
	$LSD_{0.01}$	0.12	0.06
	40	9.20 a	0.81 c
Means	60	8.34 b	1.74 b
	80	7.28 c	2.73 a
	$LSD_{0.01}$	0.13	0.07

ICD: Irrigation circuit design, L.L.L.: Lateral line length, CM_2DIS: Closed circuits with tow manifolds separated, CM_1DIS: Closed circuits with one manifold, TDIS: Traditional drip irrigation system.

The effect of DIC × LLL on PH was significant at the 1% level except between the two interactions: CM_2DIS × LLL_3 and CM_1DIS × LLL_2. The highest (9.5 m) and the lowest (6.05 m) values of PH were achieved in the interactions: CM_2DIS × LLL_1 and TDIS × LLL_3, respectively.

It is worthy to mention that the allowable drop in pressure between the maximum and minimum pressure along the lateral lines must be ≤1.1 m under turbulent flow condition. This is very necessary for drip irrigation system to be economic and water and fertilizers distribution along the lateral to be acceptable. Data, indicated that all LLL of 16 mm inside Ø under TDIS and that of 80 m in length under CM2DIS and CM_1DIS were not recommended to avoid high cost and the lower uniformity of both water and fertilizers distribution along the LLL. Therefore, for 16 mm inside Ø and 80 m long laterals, either LLL should be shorten or their inside Ø should be increased.

Data given in **Table 1** and plotted in **Figure 5** indicated that the change in friction loss took an opposite trend to that of PH. Friction loss increased with distance from lateral inlet reaching its maximum at 50% to 60% of lateral length, then it decreased again up to the lateral line end in the case of using CM_2DIS and CM_1DIS. In other words, the minimum values of friction loss existed at both the inlets and the end of the lateral lines. Reasons for this are due to the direct relation between friction loss from one side and its length and discharge from the other side.

According to the friction loss values, DIC could be ranked in the following descending order: TDIS > CM_1DIS > CM_2DIS. Differences in friction loss between any two DIC were significant at the 1% level.

The ascending order: $LLL_1 < LLL_2 < LLL_3$ illustrated the mean effect of LLL on friction loss. Differences in friction loss among LLL treatments were significant at the 1% level.

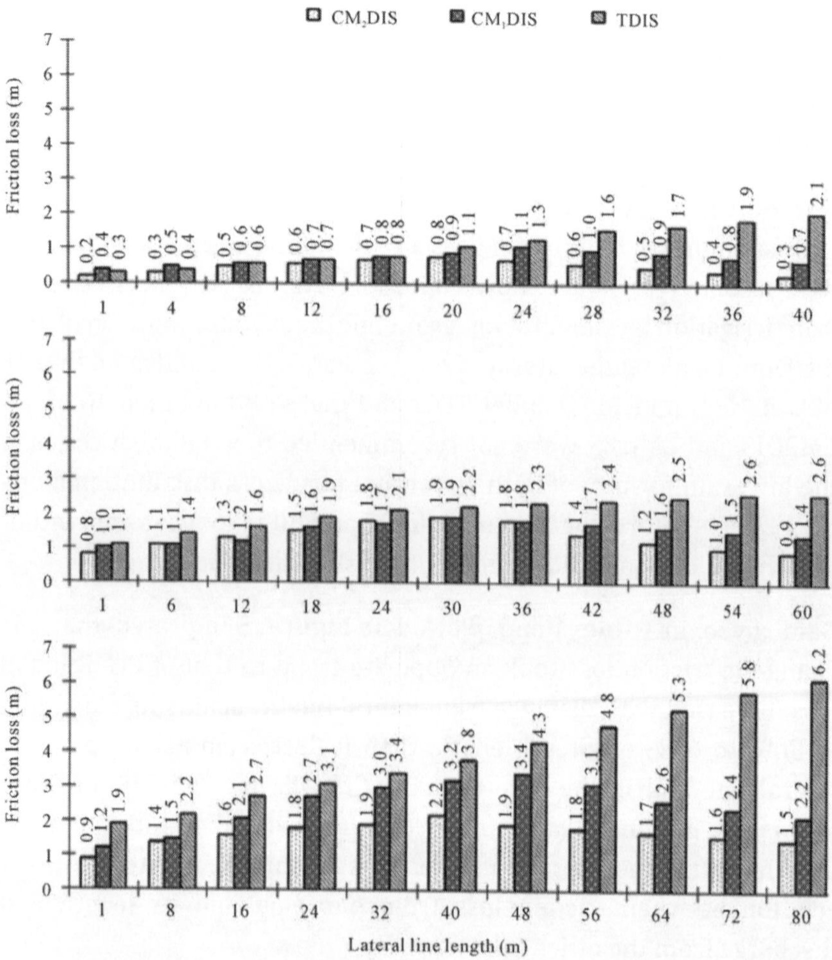

Figure 5. Effect of different irrigation circuits designs on friction loss along different lateral line lengths under (operating pressure 1.0 atm and slope = 0%).

The effect of the DIC × LLL on friction loss was significant at the 1% level. The maximum and minimum values of friction loss were obtained in the interactions: TDIS × LLL_3 and CM_2DIS × LLL_1, respectively.

As the flow rate in lateral line decreases (with respect to its length due to dripper discharges from the lateral lines) the energy gradient line will not be a straight line but a curve of an exponential type Figures 6-8. This is in agreement with [18,19]. [19] mentioned that only the total friction drop ratio (ΔH/H) affected the shape of the energy gradient lines. It is clear from Figures 6-8 that all factors affecting the ratio (ΔH/H) including DIC and LLL used also affected the shape of the energy gradient lines.

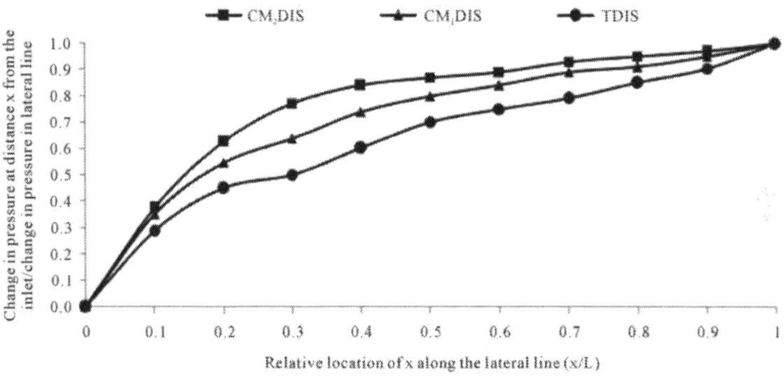

Figure 6. Dimensionless curves showing the friction drop pattern in drip lateral line under different irrigation circuits (lateral line length = 40 m, operating pressure = 1.0 atm and slope = 0%).

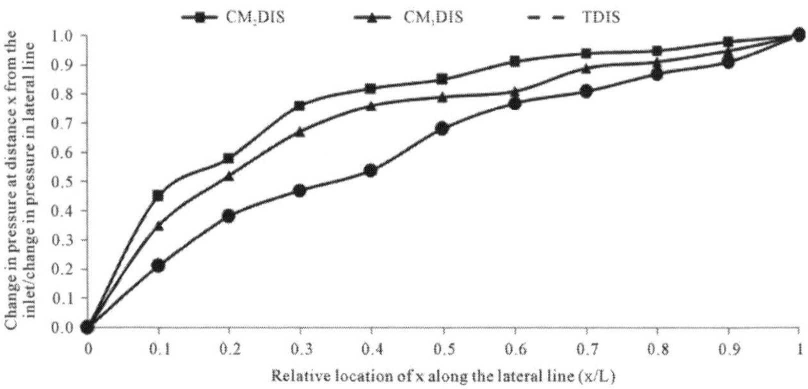

Figure 7. Dimensionless curves showing the friction drop pattern in drip lateral line under different irrigation circuits (lateral line length = 60 m, operating pressure = 1.0 atm and slope = 0%).

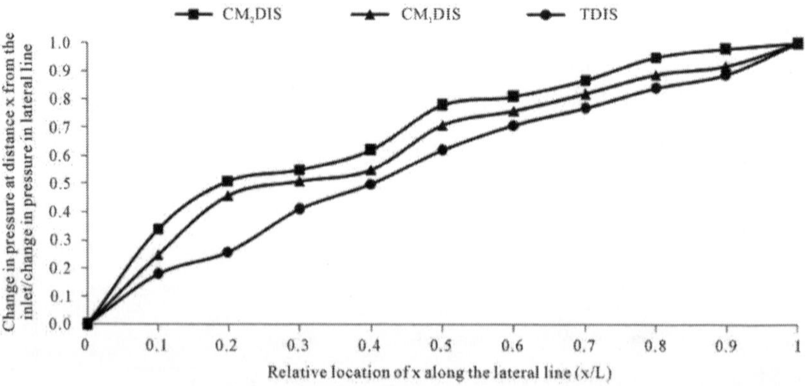

Figure 8. Dimensionless curves showing the friction drop pattern in drip lateral line under different irrigation circuits (lateral line length = 80 m, operating pressure = 1.0 atm and slope = 0%).

4. SUMMARY AND CONCLUSIONS

Closed-circuits designs were proposed as incorporating modifications to the TDIS. There were seven major conclusions from the work:

1) From LLL and PH values, DIC could be ranked in the following ascending order: TDIS < CM1DIS < CM_2DIS. The differences in PH among DIC were significant at the 1% level.

2) The depressive effects of LLL on PH could be ranked in the following ascending order: $LLL_1 < LLL_2 \leq LLL_3$. Differences in PH among LLL treatments were significant at the 1% level except that between LLL_2 and LLL_3.

3) The effects of interactions: DIC X LLL on PH were significant at the 1% level with some exceptions.

4) The highest value of PH (9.5 m) and the lowest one (6.05 m) were achieved in the interactions: $CM_2DIS \times LLL_1$ and $TDIS \times LLL_3$, respectively.

5) The shapes of the energy gradient lines were affected by DIC and LLL treatments used through their effect on ΔH/H ratio, but they took the same trend.

6) According to the FL values, DIC and LLL treatments could be ranked in the following descending orders: TDIS > CM_1DIS > CM_2DIS and LLL_1 > LLL_2 > LLL_3, respectively. The differences in FL among DIC and LLL were significant at the 1% level.

7) The effects of interaction: DIC X LLL on FL were significant at the 1% level. The maximum and minimum values of FL were obtained in the interactions: TDIS × LLL_3 and CM_2DIS × LLL_1, respectively.

Therefore, the CM_2DIS system is recommended for use where technically feasible.

REFERENCES

1. Kirnak, H., Dogan, E., Demir, S. and Yalcin, S. (2004) Determination of hydraulic performance of trickle irrigation emitters used in irrigation system in the Harran Plain. Turkish Journal of Agriculture and Forestry, 28, 223- 230.

2. Hathoot, H.M., Al-Amoud, A.I. and Mohammed, F.S. (1993) Analysis and design of trickle irrigation laterals. Journal of Irrigation and Drainage Engineering, 119, 756-767.

3. Yildirim, G. and Agiralioglu, N. (2008) Determining operating inlet pressure head incorporating uniformity parameters for multi outlet plastic pipelines. Journal of Irrigation and Drainage Engineering, 134, 341-348.

4. Deba, P.D. (2008) Characterization of drip emitters and computing distribution uniformity in a drip irrigation system at low pressure under uniform land slopes. M.Sc. Thesis, Texas A&M University.

5. Bralts, V.F., Edwards, D.M. and Wu, I.-P. (1987) Drip irrigation design and evaluation based on the statistical uniformity concept. Advances in Agronomy, 4, 67-117.

6. Gerrish, P.J., Shayya, W.H. and Bralts, V.F. (1996) An improved analysis of incorporating pipe components into the analysis of hydraulics networks. Transactions of the ASAE, 39, 1337-1343.

7. Von Bernuth, R.D. (1990) Simple and accurate friction loss equation for plastic pipes. Journal of Irrigation and Drainage Engineering, 116, 294-298.

8. Hathoot, H.M., Al-Amoud, A.I. and Mohammed, F.S. (1991) Analysis of a pipe with uniform lateral flow. Alexandria Engineering Journal, 30, C49-C54.

9. Smajstrla, A.G. and Clark, G.A. (1992) Hydraulic performance of microirrigation drip tape emitters. ASAE Paper No. 92-2057.

10. Wood, D.J. and Rayes, A.G. (1981) Reliability of algorithms for pipe network analysis. Journal of the Hydraulics Division, 107, 1145-1161.

11. Narayanan, R., Steele, D.D. and Scherer, T.F. (2000) Computer model to optimize above-ground drip irrigation systems for small areas. Applied Engineering in Agriculture, 18, 459-469.

12. Safi, B., Neyshabouri, M.R., Nazemi, A.H., Masiha, S. and Mirlatifi, S.M. (2007) Subsurface irrigation capability and effective parameters on onion yield and water use efficiency. Journal of Scientific Agricultural, 1, 41-53.

13. Netafim Irrigation, Inc. (2002) Bioline design guild. http://www.netafim.com

14. Mogazhi, H.E.M. (1998) Estimating Hazen-Williams coefficient for polyethylene pipes. Journal of Transportation Engineering, 124, 197-199.

15. Bombardelli, F.A. and Garcia, M.H. (2003) Hydraulic design of largediameter pipes. Journal of Hydraulic Engineering, 129, 839-846.

16. Hathoot, H.M., Al-Amoud, A.I. and Mohammed, F.S. (1994) The accuracy and validity of Hazen-Williams and scobey pipe friction formulas." Alexandria Engineering Journal, 33, C97-C106.

17. Steel, R.G.D and Torrie, J.H. (1980) Principles and procedures of statistics. A biometrical approach. 2nd Edition, McGraw Hill Inter Book Co, Tokyo.

18. Bazaraa, A.S. (1982) Pressurized irrigation systems sprinkler and trickle irrigation. Faculty of Engineering, Cairo University, Cairo.

19. Wu, I.P. (1992) Energy gradient line approach for direct hydraulic calculation in drip irrigation design. Irrigation Science, 13, 21-29.

6

IMPLICATIONS OF DESIGN, MANAGEMENT AND RECESSION PHASE IN DRIP IRRIGATION ON THE TOTAL DISTRIBUTION EFFICIENCY IN BLUEBERRY (VACCINIUM CORYMBOSUM L.) CROPS IN AREAS WITH HIGH SLOPES IN CONCORDIA, ARGENTINA

A. Pannunzio[1], E. A. Holzapfel[2], P. Texeira Soria[1], F. Bologna[3]

[1]University of Buenos Aires, Buenos Aires, Argentina

[2]Centro de Recursos Hídricospara la Agricultura y la Minería, Chillán, Chile

[3]Berries del Sol S.A., Colonia Ayuí, Argentina

ABSTRACT

The increase of human population generates the need to improve the efficiency of food production. A thorough planning is required following the scope of economic and sustainable development, being irrigation a basic tool, however water availability is restricted and it obliges farmers to progress increasing water productivity. Irrigation uses around 70% of total available fresh water, while irrigation water application efficiency is around 40%. Irrigation systems must follow strong criteria at the design stage to achieve high values of water productivity. Maintenance is indispensable to follow the original functioning level of those systems. At last the daily precise management of systems, following soil water potential, considering the effective rain storage at root depth of the crop and the evolution of daily evapotranspiration, preserving natural resources, are relevant to achieve low values of water footprint of this crop. In an 8 year drip irrigation system, the Uniformity Coefficient of Christiansen (UCC) measured was 95.14%; the Uniform Coefficient of the Minor Quart (UCMQ) was 93.16%. The Total Distribution Efficiency (EDT) was 95.13% when measurements finished while the irrigation systems is of. When measurements also considered the volume collected during the "recession phase in drip irrigation" and the "volume of water collected during recession phase in drip irrigation" collected at different points, EDT was 95.13%. Moreover it can be seen that when three different typical soil of the area were considered, the EDT was, 91.85%, 91.47% and 90.30% respectively, according with different water storage capacity of each soil. The Total Distribution Efficient is a strong method, to evaluate the design and management of drip irrigation systems, under different design criteria, management practices and maintenance of the systems. Water footprint in a blueberry (Vaccinium corymbosum L.) crop with drip and sprinkler anti-frost system, were measure and values obtained were 846, 310, 223, 212, 172 and 218 liters per kg of fresh fruit in the period 2010-2015. The UCC and the UCMQ reflects properly the irrigation design, while the EDT reflects irrigation design, management and maintenance. Water footprint is at last the strong tool to evaluate design and operation of the irrigation system and crop management.

Keywords: Recession Phase in Irrigation, Water Management, Drip Irrigation, Irrigation Design Criteria, Drip Irrigation Management, Blueberry Crop, Snow Chaser Variety, Irrigation Uniformity, Water Footprint

1. INTRODUCTION

Water is vital, multifunctional and scare resource, and these characteristics generate a strong competition among water users. A thorough planning is required following the scope of economic and sustainable development. During the seventies, the deficient management of water resources became relevant and the paradigm of Integrated Management of Hydric Resources [1] was suggested.

The increase of human population generates the need to improve the efficiency of food production. At this point, irrigation is a basic tool, however water availability is restricted and it obliges farmers to progress in water productivity. Irrigation uses near 70% of total fresh water available and the water application efficiency is around 40% [2] .

The use of indicators is basic to reach sustainability [3] and to improve irrigation water management [4] at the design and managing stage of irrigation systems [2] . The water footprint is a good indicator of efficiency of water used in different processes, and it includes direct and indirect use of water to produce goods or services during a certain period [5] .

Water footprint concept was proposed in the XXI century by Hoekstra [6] . This concept divides water use considering its source and contamination associated with productive process. In 1993, Allan introduced virtual water concept, measuring the water contained in each product and the water used during the process [7] . Both concepts answer the requirements of quantification of this water use, however water footprint implies volume of water used and virtual water implies the flow of the water as a net balance of the water of a country. Chapagainand Hoekstra estimated water footprint for several countries for the period between 1997-2001 [8] . Currently water footprint is classified as: 1) Blue water that is related with the water use from superficial and ground water sources [9] ; 2) Green water is related with the rain water used by the crop; and 3) Grey water is the contaminated water used during the process [10] . Considering the difficulties of measuring grey water, this water is not thoroughly studied as blue and green waters. The differentiation of three types of water is crucial due to environmental implications, thus requiring several management policies [5] .

Several authors have considered water footprint in agriculture. Mekkonen and Hoekstra [11] quantified blue, green and grey water for global production of 126 crops. Ridoutt et al. [12] calculated water footprint for mango. Deuret el al. [14] calculated water footprint for kiwi crop. Herath et al. [13] studied grey water of potato crop in New Zeeland. In 2013, the Instituto de Investigaciones Agropecuarias de Chile edited the book "Determinación de la Huella del Agua y estrategias de manejo de los Recursos Hídricos", for several crops including blueberry crop.

The general values for blueberries water footprint are around 341, 334 and 170 L/kg for green, blue and grey footprints, resulting in a total of 845 L/kg [15] . These values are consistent with those obtained in Chile between 400 and 800 L/kg [16] .

As an example of the importance of water footprint as an indicator, the ISO 14.046 Standard quantifies water environmental impacts and the possibilities of reducing its effects. In 2014, the Autoridad del Agua de la Provincia de Buenos Aires incorporated water footprint to establish water cost for Argentina farmers. Water footprints for agricultural production; have been studied by various authors in Argentina [17] with studies of water footprint in rice production in Entre Ríos [18] .

The knowledge of water footprint in blueberry (Vaccinium corymbosum L.) crop is an important issue to plan efficient water use, improving productivity, sustainability and competitiveness of irrigated crops [1] , [17] . Blueberry in Argentina is destined to the North Hemisphere markets, taking advantage of commercial window during the period of September to December. To produce in September, early growing varieties are required. These varieties need anti-frost protection system during the winter, because some frost occurs in Concordia area of Argentina while plant sensitivity to low temperature is high. Sprinkler, Micro-sprinkler and Mini-sprinkler solid set systems are used to avoid frost damages during the nights in which temperatures below zero happens.

Two main regions produce more than 80% of the Argentinean production of blueberries: Concordia region, in Entre Rios province in Northeast and the other in the Northwest area of Argentina, in Tucuman and Salta provinces.

This paper discusses water footprint research in blueberries under drip irrigation in Concordia, Argentina.

2. MATERIAL AND METHODS

The trail was developed in a farm located at Colonia Ayuí, Concordia, Entre Ríos, Argentina, with the following coordinates: 31°11'24"S, 58°02'54"W. There is a commercial farm called Berries del Sol S.A., with 30 has of Southern High bush Blueberry varieties (Vaccinium corymbosum, L.). Rainfalls in the area are between 800 and 1700 mm annually all spread around the year. There are sandy soils in the area over a clay layer. Slopes are around 1.7%, with high erosion risk, when combined with the mentioned rainfall amounts. During winter there are frost events, producing some damages for the early crop varieties, as Snow chaser variety and Emerald. This implies the requirement of frost protection systems. Water for irrigation in this area can be obtained from deep wells, these wells can produce usually 300 $m^3 \cdot h^{-1}$. Some farms also built some dams, with storage capacity to provide water considering the requirements of a sprinkler frost protection system, with a precipitation rate of 3.5 $mm \cdot h^{-1}$ or 35 $m^3 \cdot h^{-1}$. Those systems worked in average 10 hours per night and 30 nights during winter and early spring.

The relation between water potential of the soils and water content was determined in laboratory to calculate the irrigation dose in order to keep the expected water potential to maximized yields.

Reference water evapotranspiration, was calculated following Penman-Monteith method using CROPWAT program of Food and Agriculture Organization of the United Nations. Crop coefficient to obtain the evapotranspiration of blueberries was calculated following Holzapfel [19] model, considering shaded area of the crop.

Effective rainfall, was calculated with the Soil and Conservation Service of United States Department of Agriculture method that calculate with accuracy precision the water stored in the root depth according with water monthly evapotranspiration and water storage capacity of the soils in the mentioned rooting depth.

The uniformity of the system was measured with the Uniformity Coefficient of Christiansen [15] and by the Uniform coefficient of the minor quarter.

The global efficiency of the drip irrigation system was calculated applying the Total Distribution Efficiency [20] coefficient.

At last water footprint of blueberries was calculated from 2010 to 2015, taking in account that the first harvest season was 2010, so water was basically used that year for plant growing, while in the following years it was also used for fruit production an also to produce fruit, reason why water footprint values decrease dramatically from 2011 to the current years, beyond the strong management efforts to increase water productivity.

3. RESULTS AND DISCUSSION

3.1. Climatic Characterization of Concordia

The study area is located within 31°11'24"S, 58°02'54"W at 40 meters above sea level in "Berries del Sol" farm in Colonia Ayuí, Concordia, Entre Ríos province. The Concordia climate is a temperate humid, with medium temperatures between 17°C to 20°C, and rainfall in the area varies from 800 to 1700 mm/year spread all over the year [8] . These values of rainfall are not enough to meet demands of water use for blueberry crop, considering the sandy soil of the area and the shallow root system of the crop (Table 1). In this region, "El Niño" phenomenon is associated with water excess during November while dry seasons can be expected in the other periods [21] .

3.2. Soil Water Potential Characterization and Its Importance to Determine Water Requirements

Concordia region is just aside the Uruguay River, a river with a flow of 5000 $m^3 \cdot s^{-1}$ limiting with Uruguay country. The west border of Entre Rios province is the Parana river with 15,000 $m^3 \cdot s^{-1}$. Bothe are affluent of the Rio de la Plata river. Sandy soils over strong clay subsoil can be found in some spotted areas, in which blueberry crop is cultivated. Main roots of the crop were found up to 20 cm, where soils texture consisted of: clay 9%, silt 10.5%, sand 80.5%, pH of ≈5, salinity of 0.3 $dS \cdot m^{-1}$ and cation exchange capacity (CEC) of 8 meq/100 g.

For research on crop production, it is important to know the crop response to different water potentials at which water is retained in the soil, determining the water potential threshold of a crop to maximize yields or increase water productivity.

To determine these values, soil samples from several plots were tested in laboratory. Water content was determined at potentials: Of 10, 20, 33 and 125 kPa. With these values and for zero to twenty centimeters depth (root activity zone of blueberries), a curve was constructed at different water potentials for three representative soils of the area (Figure 1).

Table 1. Evapotranspiration with Penman-Monteith method (ETo) and climatic data: Concordia region.

Month	Min Temp	Max Temp	Humidity	Wind	Sunshine	Radiation	ETo
	Celcius	Celcius	%	m/s	Daily hours	MJ/m²/day	mm/day
January	8.4	42.2	63	3.6	8.9	24.6	8.69
February	7.0	41.6	68	3.4	7.7	21.5	7.75
March	4.7	39.9	72	3.2	7.7	19.0	6.71
April	1.5	35.0	75	3.0	6.1	13.8	4.89
May	−1.3	31.8	78	3.1	5.9	11.0	3.96
June	−5.3	29.0	80	3.2	4.6	8.6	3.42
July	−3.9	31.6	78	3.5	5.1	9.5	4.04
August	2.2	32.4	74	3.6	5.7	12.2	4.7
September	−3.0	34.4	78	3.8	6.5	16.1	5.92
October	0.9	37.8	72	3.9	7.8	20.7	7.31
November	2.3	39.0	69	3.7	8.5	23.6	7.96
December	5.4	41.0	64	3.6	9.3	25.6	8.61
Average	1.2	36.3	72	3.5	7.0	17.2	6.16

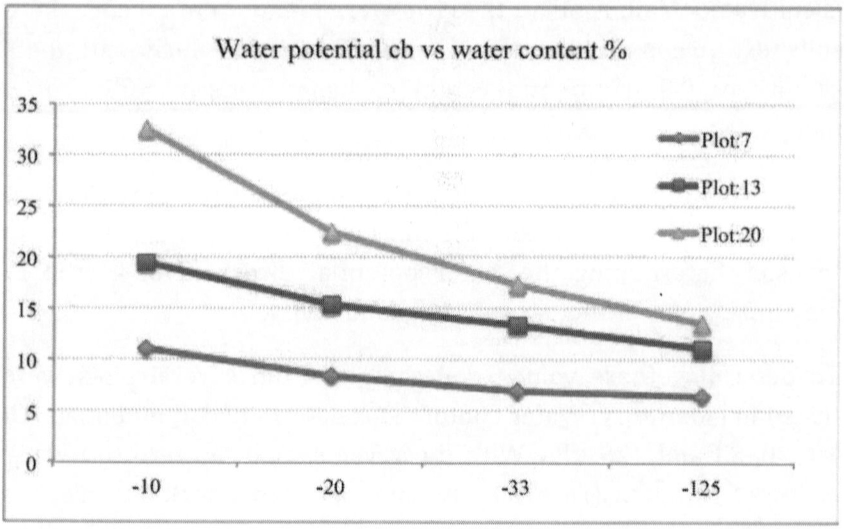

Figure 1. Soil moisture retention curve for two layers soils (Y-axis: % moisture and X-axis: Tension in cb), for three different plots, for the 0 to 20 cm layer.

Blueberry crop is very sensitive to water stress because of superficial root system [9] .Around 80% of the water is extracted from first 20 cm soil depth with significant root activity, even though roots can extend up to 40 cm. Therefore, crop water requirements at

20 cm depth were studied in this research. For analyzing the water potential interval for irrigation scheduling, the range of 10 - 20 kPa was considered [22] , because water potential greater than 20 kPa implies decrease in yield [23] . Based on these results, a strict range of water potential must be followed to obtain high yields.

For this range of the water content of those sandy soils, only a small quantity of water should be applied in each irrigation procedure, to maintain the expected water potential, avoiding deep percolation and also water stress. In this farm, plants were spaced at 0.85 m with row spacing of 3.5 m. The wetted area per plant was observed to be 0.45 $m^2 \cdot plant^{-1}$, while the drip system is working.

3.3. Water Requirements for Blueberry Crop in Concordia

The estimation of water requirements is a key factor to design and operate irrigation system. Evapotranspiration demand of crop was estimated using the Penman-Monteith formula, crop coefficient (kc) and crop coverage factor [11] .

The Penman-Monteith includes all parameters that govern energy exchange and corresponding latent heat flux (evapotranspiration, ET) from uniform expanses of vegetation. Most of the parameters are measured or can be readily calculated from weather data. The equation can be utilized for the direct calculation of any crop evapotranspiration as the surface and aerodynamic resistances are crop specific.

$$ETc = ETo * Kc * Fc \quad (1)$$

where, ETo is the reference evapotranspiration, Kc is the crop coefficient and Fc is the coverage factor depending on the percentage of shaded area of the crop.

ETo is calculated using empirical formulas, Class A pan evaporating, lysimeter measurements. In this study, Penman-Monteith formula was applied using CROPWAT 8.0 with data from Table 1. The effective rainfall is required to know the effective water provided by rainfall, and moisture storage in the root zone depth. One must also consider water percolation and erosion potential. To calculate the effective rainfall, the USDA-Soil Conservation Service method was used. This method considers monthly rainfall and monthly crop evaporation crop (Table 2).

The correct determination of the crop coefficient (Kc) is basic to irrigation requirements and to manage irrigation systems scheduling [22] . Several researchers have indicated Kc value of blueberry crop between 0.2 - 0.97 for a crop of one to three years old [23] , while others indicate Kc of 0.2 to 1.1 for the same crop [24] . Number of

Table 2. The effective moisture storage (mm) in the soil in mm.

	2010	2011	2012	2013	2014	2015	Average
January	31	101	15	31	85	96	58
February	64	100	100	82	100	29	80
March	27	50	60	50	60	14	44
April	32	38	0	26	48	8	29
May	6	20	16	20	20	20	17
June	16	12	0	0	20	20	11
July	12	20	8	14	18	0	15
August	24	18	36	13	6	40	22
September	60	45	0	14	21	40	30
October	0	22	0	29	58	32	24
November	120	9	0	38	85	96	47
December	62	76	121	0	140	48	73
Average	455	511	356	317	660	442	451

plants per hectare must also be considered [6] . In this study, plants per hectare were 3465. Research studies have shown high crop yield at 100% of water demand, while applying 150% of the requirements does not give high yield [25] , [26] .

3.4. Timing Options in Drip Irrigation with the Measured Parameters

Three tables (Tables 3-5) were elaborated considering the upper mentioned parameters, to obtain the number of daily irrigations required to fill water demand of the crop during peak, for blueberry crops under the conditions of Concordia region. Is important to see that three plots in the same farm, requires different timing irrigations in- tervals, according with the water potential values of each soil. That means that a proper management of the drip irrigation system must operate each type of soil in different way, replacing the exact required volume of water in a different irrigation interval.

For plant number seven, 2.43 liters must be filled 5 times per day. For plot number thirteen, 3.78 liters must be filled 3 times per day, and for plot 20 with a wider range of water content for the same water potential range, 8.91 liters must be given 1.3 times per day.

Table 3. Irrigation timing for plot number 7.

Plot number 7:		
Peak evapotranpiration	8	mm·day^{-1}
Wetted area per plant:	0.45	m^2
Relevant depth of roots for blueberry crops:	0.2	m
Wetted volume of soil:	0.09	m^3
Water content at 10 cb:	11.2	%
Water content at 20 cb:	8.5	%
Available percentage of water range for that interval:	2.7	%
Volume of water to be replaced:	2.43	liters·plant^{-1}
Flow per hour for 1 liter hour drippers each 30 cm, two drip lines per row:	6	liters·hour^{-1}
Dally irrigation time to supply water requirements:	24.3	minutes
Shaded area per plant:	1.53	m^2
Daily demand for peak according with shaded area:	12.24	liters·plant^{-1}·day^{-1}
Minimum operations per day to supply water demand:	5	Daily operations

Table 4. Irrigation timing for plot number 13.

Plot number 13:		
Peak evapotranpiration	8	$mm \cdot day^{-1}$
Wetted area per plant:	0.45	m^2
Relevant depth of roots for blueberry crops:	0.2	m
Wetted volume of soil:	0.09	m^3
Water content at 10 cb:	19.5	%
Water content at 20 cb:	15.3	%
Available percentage of water range for that interval:	4.2	%
Volume of water to be replaced:	3.78	$liters \cdot plant^{-1}$
Flow per hour for 1 liter hour drippers each 30 cm, two drip lines per row:	6	$liters \cdot hour^{-1}$
Dally irrigation time to supply water requirements:	37.8	minutes
Shaded area per plant:	1.53	m^2
Daily demand for peak according with shaded area:	12.24	$liters \cdot plant^{-1} \cdot day^{-1}$
Minimum operations per day to supply water demand:	3	Daily operations

Table 5. Irrigation timing for plot number 20.

Plot number 20:		
Peak evapotranpiration	8	$mm \cdot day^{-1}$
Wetted area per plant:	0.45	m^2
Relevant depth of roots for blueberry crops:	0.2	m
Wetted volume of soil:	0.09	m^3
Water content at 10 cb:	32.4	%
Water content at 20 cb:	22.5	%
Available percentage of water range for that interval:	9.9	%
Volume of water to be replaced:	8.91	$liters \cdot plant^{-1}$
Flow per hour for 1 liter hour drippers each 30 cm, two drip lines per row:	6	$liters \cdot hour^{-1}$
Dally irrigation time to supply water requirements:	89.1	minutes
Shaded area per plant:	1.53	m^2
Daily demand for peak according with shaded area:	12.24	$liters \cdot plant^{-1} \cdot day^{-1}$
Minimum operations per day to supply water demand:	1.3	Daily operation

3.5. Design and Evaluation of Drip Irrigation System

English et al. [27] mentioned that irrigated agriculture will need to provide two-thirds of the increase in food to feed the growing population. The drip irrigation system was designed considering two laterals of drippers at a dripper spacing of 30 cm and drippers with 1 liter per hour per emitter. Mini-sprinkler irrigation system is commonly used as anti-frost system in this area as mentioned before. Wells at the site can produce up to 300 $m^3 \cdot h^{-1}$ for anti-frost irrigation purpose. The pH of the water is 6.5 with electrical conductivity of 0.14 dSm^{-1}. The design and operation of the system are relevant to improve water management and the economic profitability of irrigated agriculture [1] . The Uniformity Coefficient of Christiansen [15] is a statistical coefficient to show the dispersion between all values and the average value.

$$UCC(\%) = \left[1 - \frac{\sum_{i=1}^{n} |x_i - \bar{x}|}{n\bar{x}} \right] \times 100 \qquad (2)$$

where, x_i is the emitter flow ($l\ h^{-1}$), $\bar{x}$ is the average flow of the evaluated emitters ($l\ h^{-1}$), and n is the number of emitters.

The UCC coefficient of 95.13% was obtained for the study areas. The Uniform coefficient of the minor quarter (UCMQ) is given below:

$$UCMQ = \frac{x_{25}}{x} \qquad (3)$$

where, x_x is the average flow of the evaluated emitters; and x_{25} is the average flow emitted by the 25% of the emitters of the minimum flow.

The UCMQ for the same emitters was 93.69%.

$$EDT = \left[1 - \frac{\sum_{i=1}^{n} |x_i - X_r|}{n * X_r} \right] \times 100 \qquad (4)$$

where, EDT is the total distribution efficiency; x_i is the water infiltrated for i point (mm); X_r is the crop water need (mm); and n is the number of observations according to Holzapfel model [20] .

The total distribution efficiency shows the way in which water distribution is compared with water requirements, as well as the soil water holding capacity in the extracted root zone. This analysis is focused in the variations between water required and water store in the root zone in each point.

The EDT considers all the water irrigated by the systems during the application time plus the water during "the recession phase" concept basically used in surface irrigation, but also very important, in drip systems, in which while following tightly soil water potential, demands several daily irrigation operations, like pulse irrigation. With several irrigations per day, the remaining volume of water during the irrigation phase will be drained of the system, according with hydraulic design, type of drippers, topographic shape of each plot, etc.

The value obtained of EDT, in plots 7, 13 and 20, were: 91.85%, 91.47% and 90.30%. This high efficient values shows that even thought there are very high, there is a strong relation between irrigation efficiency and criteria used during the design and management system, according, with soils, crops, slopes, water quality, etc.

3.6. Blueberry (Vaccinium corymbosum L.) Yield (VAR. Snowchaser): Concordia, entre RÍos, Argentina

To quantify water productivity we measure the yield of a Snow chaser blueberry variety, planted in the farm during 2010. The average yields obtained per plant from 2010 to 2015 are shown inFigure 2.

3.7. Estimation of Water Footprint for Blueberry (Vaccinium corymbosum L.) (Snowchaser VAR.) in Concordia, entre RÍos, Argentina

Irrigation sustainability in blueberry crops is an important issue [28] . Blue water is the water applied by the drip irrigation system. The green water is the water received by rainfalls and represents effectively storage in the soil volume explored by roots. Grey water is the water used by the minisprinkler frost protection irrigation system. The effective rain was calculated following the procedure of the Soil Conservation Service of the United States Agriculture Department. In the case of the grey water, authors considered the water applied by the sprinkler anti-frost irrigation system. Considering different components of the water footprint and the yield obtained, water footprint for period 2010 to 2015 of Snow chaser variety under Concordia conditions is shown in Figure 3.

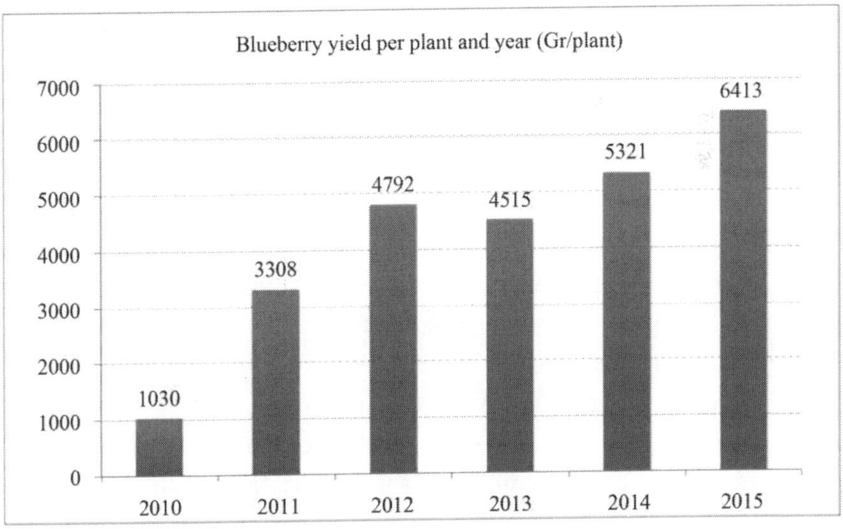

Figure 2. Blueberry yield: Grams per plant/year.

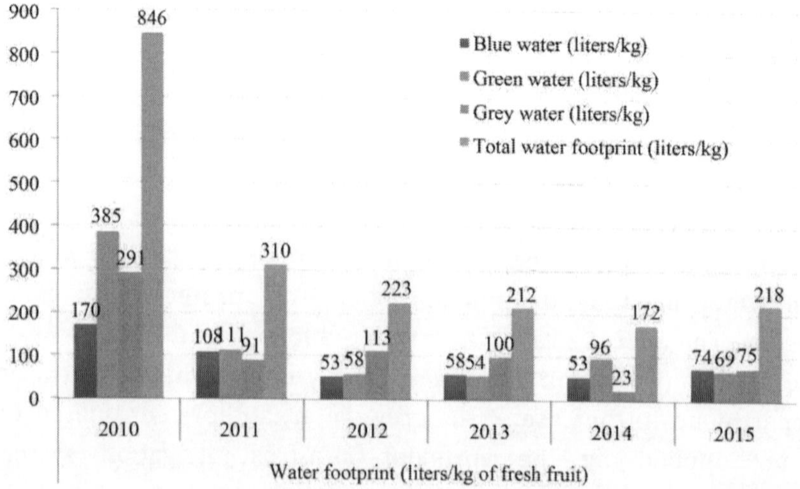

Figure 3. Water footprint: Liters per kg of blue, green and grey water per kg of fruits

4. CONCLUSION

Food requirements for an increasing population encourage agriculture to produce more food, taking in account that rain fed crops are the protagonists of producing additional food. Available land for crops are a limited and agriculture is using around 70% of available fresh water. So great efforts must be done, to produce food taking care of water productivity, always following sustainable principles. In that way some tools can be used to reach the scope. First of all a proper design of irrigation systems is imperative to install efficient irrigation systems. Is essential a daily measure of crop requirements and water potential of the soil in order to determine the exact moment, for each irrigation operation. Management also implies the correct maintenance of drippers, filters, valves and all the components of the system. Farmers can be advised to check their systems using: The Uniformity Coefficient of Christiansen (UCC), the Uniform Coefficient of the Minor Quart (UCMQ) and the Total Distribution Efficiency (EDT). At last water footprint is a great tool to measure water productivity and to compare different systems, and different crops, while water is a scarce resource. Water footprint involves aspects related to the basin, well ruled and legal monitoring of rights and obligations of all social actors; and it can give a fair frame to reach better results for the community without compromising environment. In an 8 year drip irrigation system in a blueberry crop in Concordia, Entre Ríos, Argentina, the Uniformity Coefficient of Christiansen (UCC) measured was 95.14%, the Uniform Coefficient of the Minor Quart (UCMQ) was 93.16%. The Total Distribution Efficiency (EDT) was 95.13% when measurements finished while the irrigation systems is switch of. When measurements also considered the volume collected during the "recession phase in drip irrigation" and the "volume of water collected during recession phase in drip irrigation" collected at different points, EDT was 95.13%. Moreover it can be seen that when three different typical soil of the area were considered, the EDT was, 91.85%, 91.47% and 90.30% respectively, according with different water storage capacity of each soil. The Total Distribution Efficient is a strong method, to evaluate the design and management of drip irrigation systems, under different design criteria, management practices and maintenance of the systems. Water footprint was measured in a blueberry crop and values obtain were 846, 310, 223, 212, 172 and 218 liters per kg of fresh fruit in the period 2010-2015.

REFERENCES

1. Goyal, M.R., Ed. (2015) Design and Management of Irrigation Systems: Chile. Título del libro: Research Advances in Sustainable Microirrigation, Sustainable Microirrigation, Principles and Practices. International Standard Book Number 13-1-978-77188-016-9 (Hardcover), Apple Academic Press, Toronto, Chapter 8, 147-160.

2. Hoekstra, A., Chapagain, A., Aldaya, M., Maite, M. and Mesfin, M. (2009) Water Footprint Manual. http://waterfootprint.org/media/downloads/WaterFootprintManual2009.pdf

3. Altieri, M. and Nicholls, C. (2005) A Rapid, Farmer-Friendly Agro Ecological Method to Estimate Soil Quality and Crop Health in Vineyard Systems. Agroecology and the Search for a Truly Sustainable Agriculture. PNUMA, 277, 290.

4. Cañas, J., Jiménez, M., Arellano, R. and Moreno-Pérez, M. (2010) Improving Water Application Using Irrigation Indicators (Mejora de la gestión del agua de riego mediante el uso de indicadores de riego). Revista Universidad Nacional de Cuyo, 42, 107-124.

5. Pannunzio A., Holzapfel, E., Texeira P. and Bologna, F. (2016) Water Footprint in Cultivated Blueberries with Drip Irrigation Systems in Concordia, Argentina. In: Goyal, M.R., Ed., Advances in Irrigation Management of Horticultural Crops: An Overview, Apple Academic Press, Toronto, in Press.

6. Hoekstra, A. and Hung, Q. (2002) Virtual Water Trade: A Quantification of Virtual Waterflows between Nations in a Relation to International Crop Trade. Value of Water Research Report Series 7.

7. Allan, J. (1998) Watersheds Explaining the Absence of Armed Conflict over Water in the Middle East. MERIA— Middle East Review of International Affairs, 2, 1-3.

8. Chapagain, A. and Hoekstra, A. (2004) Water Footprints of Nations. In: Value of Water Research. Report Series No. 16. UNESCO-IHE. Delft, the Netherlands. 11, UNEASCO—IHE, Delft, The Netherlands.

9. Rodríguez Casado, R., Garrido Colmenero, A., Llamas Madurga, M. and Varela Ortega, C. (2008) Hydrological Foot Print of Spanish Agriculture. Reportes de Agua Virtual. Fundación Marcelino Botín.

10. Falkenmark, M. (2003) Freshwater as Shared between Society and Ecosystems: From Divided Approaches to

Integrated Challenges. Philosophical Transactions of the Royal Society of London (Series B, Biological Science), 358, 2037-2049.

11. Holzapfel, E., Pannunzio, A. and Lorite, I. (2014) Design and Management of Irrigation Systems. In: Goyal, M., Ed., Research Advances in Sustainable Microirrigation Principle and Practices, Volume 1, Apple Academic Press Inc., New Jersey.

12. Pritchard, R., Haman, D., Smajstrla, A. and Lyrene, P. (1993) Water Use and Irrigation Scheduling of Young Blueberries. Proceedings of the Florida State Horticultural Society, Publication 106, Florida, 147-150.

13. Herath, I., Clothier, B., Green, S., Horne, D., Singh, R., Marsh, A., Buchanan, A. and Burgess, R. (2012) Measuring the Grey-Water Footprint of Potato Production. http://www.massey.ac.nz/~flrc/workshops/12/Manuscripts/Herath_2012.pdf

14. Deuret, M., Green, S., Clothier, B. and Mowat, A. (2011) Can Product Water Footprints Indicate the Hydrological Impact of Primary Production?—A Case Study of New Zealand Kiwifruit. Journal of Hydrology, 408, 246-256.

15. Mekkonen, M. and Hoekstra, A. (2010) A Global and High-Resolution Assessment of the Green, Blue and Grey Water Footprint of Wheat. Hydrology and Earth Systems Sciences, 14, 1259-1276.

16. Cifuentes, H. and Merino, F. (2013) Hydraulic Footprints to Determine Water Use and to Manage Water Resources. Book Series 50, 210. (Spanish)

17. Pannunzio, A. and Texeira-Soria, P. (2010) Fertigation Trial in Four Blueberries Varieties in nOrthern Argentina. 28th International Horticultural Congress, Lisboa-Portugal, 22-27 August 2010, 280.

18. Bryla, D. and Strik, B. (2005) Water Requirements of High Bush Blueberry Planted at Normal and High Densityspacings. HortScience, 40, 1058-1059.

19. Holzapfel, E.A. (2002) Irrigation in Cranberry. International Congress on Vaccinium corymbosum, Universidad de Concepción, Chillan, Chile.

20. Holzapfel, E.A. (1984) Selection and Design of Surface Irrigation Methods. PhD Thesis, University of California, Davis.

21. Heinzenknecht, G. (2005) Impact of "El Niño or La Niña" on Rainfall. Report by Office of Risks and Agriculture, Ministry of Agriculture, Ganadería y Pesca. http://www.ora.gov.ar /informes/enso_precipitaciones.pdf

22. Haman, R., Pritchard, T., Smajstrla, A., Zazueta, F. and Lyrene, P. (1997) Evapotranspiration and Crop Coefficients for Young Blueberries in Florida. Applied Engineering in Agriculture, 13, 209-216.

23. Pannunzio, A., Vilella, F., Texeira-Soria, P. and Premuzik, Z. (2011) Impact of Irrigation Systems in Blueberries. Revista Brasileira de Engenharia Agrícola e Ambiental, 15, 3-8.

24. Bryla, D. and Strik B. (2007) Effects of Cultivar and Plant Spacing on the Seasonal Water Requirements of High Bush Blueberry. HortScience, 132, 270-277.

25. Bryla, D., Yorgey, B. and Shireman, A. (2009) Irrigation Management Effects on Yield Andfruit Quality of Highbush Blueberry. Acta Horticulturae, 810, 649-656.

26. Bryla, D. (2008) Water Requirements of Young Blueberry Plants Irrigated by Sprinklers, Microsprays and Drip. Proceedings on Irrigation of Horticultural Crops, ISHS Acta Horticulturae 792, Corvalis, 135-139.

27. English, M.J., Solomon, K.H. and Hoffman, G.J. (2002) A Paradigm Shift in Irrigation Management. Journal of Irrigation and Drainage Engineering, 128, 267-277.

28. Pannunzio, A. (2010) Efectos de sustentabilidad de los sistemas de riego en arándanos. Orientación, Buenos Aires, 86 p.

7

EFFECT OF CLOSED CIRCUITS DRIP IRRIGATION SYSTEM AND LATERAL LINES LENGTH ON GROWTH, YIELD, QUALITY AND WATER USE EFFICIENCY OF SOYBEAN CROP

Mohamed Yousif Tayel, Hani Abdel-Ghani Mansour*, Sabreen Khalil Pibars

Water Relation and Field Irrigation Department, National Research Center, Giza, Egypt

ABSTRACT

Field experiment was conducted for one growing season (2012) in clay loam soil at the Experimental Farm of Faculty of Agriculture, Southern Illinois University at Carbondale (SIUC), USA on soybean crop to study the effect of: 1) the closed drip irrigation system: closed circuits with one and two a manifolds for lateral lines (CM1DIS; CM2DIS) and traditional drip irrigation system (TDIS) as a control; and 2) lateral lines length (LLL): LLL1, LLL2 and LLL3 (40, 60, 80 m) on soybean growth, yield, oil, protein content and water use efficiency. Plants were drip irrigated every 4 days. N, K_2O and P_2O_5 fertilizers were added via irrigation water. Data obtained could be outlined as follows: 1) According to the mean values of soybean crop growth (leaf area; plant height), yield (grain and straw), both oil and protein content and water use efficiency, the treatment used could be ranked in the following ascending orders: TDIS < CM1DIS < CM2DIS and LLL3 < LLL2 < LLL1; 2) Differences in the means of the studied data among treatments used were significant at the 1% level; 3) The effects of the DIC × LLL on the data obtained were significant at the 1 % level; and 4) The highest values of the obtained data and the lowest ones were achieved in the following interactions: CM2DIS × LLL1; TDIS × LLL3, respectively.

Keywords: Closed Circuits; Drip Irrigation; Lateral Lines; Soybean; WUE

1. INTRODUCTION

Soybean is one of the most important world crops. It is grown for oil and protein [1]. It had been mentioned that the present world production is about 6.2 million tons of seed over 45 million ha. According to [1] the growth periods most sensitive to water deficit of soybean are the last part of the flowering stage and the early part of the yield formation (pod development). When water supply is limited, saving in water can be made by reducing the supply during the vegetative period and near crop maturity. [2] mentioned that irrigation is an important and an increasingly common practice in Georgia and other southeastern states for soybean production, as shown by an expansion in irrigated acreage from almost 9000 ha in 2000 to more than 40,000 in 2008. Soybean yield had been reported as low as 807 kg·ha−1 in 1980 and as high as 2220 kg·ha−1 in 2003 [3]. This large difference are mainly due to droughts, evidencing the need for supplemental irrigation in Georgia, despite humid climate. Several studies conducted for a wide range of environments have demonstrated that soybean yield increases with irrigation [4-7].

Closed drip irrigation circuits have used in attempts to overcome the drop in pressure at the end of the lateral line of drip irrigation system. [8-13] carried out laboratory and field experiments to study the effects of: Closed drip irrigation circuit with one manifold for lateral lines (CM1DIS), with two manifolds for lateral lines (CM2-DIS); traditional drip irrigation system as a control (TDIS) and lateral lines and length (LLL) of 40, 60, 80 m (LLL1, LLL2, LLL3) and their interactions on some hydraulic characteristics of the irrigation system, corn yield, water, fertilizer use efficiency per unit of irrigation water and fertilizer used. Their data could be summarized in the following: 1) Relative to TDIS, both CM2DIS and CM1DIS improved the studied hydraulic characteristics (pressure head, friction loss, flow velocity, lateral discharge, uniformity coefficient; coefficient of variation), corn yield and use efficiency of both water and fertilizer; 2) The mean effects of both DIC and LLL treatments on the studied parameters were significant at the 1%; 3) The effect of DIC × LLL on the parameters under investigation were significant at the 1% level; and 4) concerning the improving effect on the investigated parameters, the treatments could be stated in the following increasing orders: TDIS < CM1DIS < CM2DIS and LLL3 < LLL2 < LLL1.

2. MATERIAL AND METHODS

The field experiments design was split in randomized complete block design with three replicates. Laboratory tests carried out using three irrigation lateral lines L1, L2; L3 (40, 60, 80 m) and the following three drip irrigation circuits (DIC): 1) One manifold for lateral lines or closed circuits with one manifold of drip irrigation system (CM1DIS); 2) Closed circuits with two manifolds for lateral lines (CM2DIS); and 3) traditional drip irrigation system (TDIS) as a control.

Tables 1-3 indicated there some (physical and chemical) characteristics of there the location and irrigation water, respectively.

Irrigation networks include the following components are: 1) Control head: It was located at the water source supply. It consists of centrifugal pump 3"/3", driven by electric engine (pump discharge of 80 m^3/h and 40 m lift), sand media filter 48" (two tanks), screen filter 2" (120 mesh), back flow prevention device, pressure regulator, pressure gauges, flow-meter, control valves and chemical Injection; 2) Main line: PVC pipes of 75mm in (ID) Ø to convey the water from the source to the main control points in the field; 3) Sub-main lines: PVC pipes of 75 mm in (ID) Ø were connected to with the main line through a control unit consists of a 2" ball valve and pressure gauges; 4) Manifold lines: PVC pipes of 50 mm in (ID) Ø were connected to the sub main line through control valves 1.5"; 5) Lateral lines: PE tubes of 16 mm in (ID) Ø were connected to the manifolds through beginnings stalled on manifolds lines; 6) Emitters: These emitters (GR) built in PE tubes 16 mm in (ID) Ø, emitter discharge of 4 lh-1 at 1 atm, Operating pressure and 30 cm spacing in-between. The components of closed circuits of the drip system include, supply lines, control valves, supply and return manifolds, drip lateral lines, emitters, check valves and air relief valves/vacuum breakers.

2.1. Irrigation Scheduling

Intervals of irrigation (I) in day were calculated using the following equations:

$$I = d/ETc \quad (1)$$

where: d = net water depth applied per each irrigation (mm), and ETc = crop evapotranspiration (mm/day).

Table 1. Some physical properties of the soil*.

Sample depth, cm	Particle Size Distribution, %				Texture class	θ% (w/w) at		
	C. Sand	F. Sand	Silt	Clay		F.C.	W.P.	A.W.
0 - 15	3.4	29.6	39.5	27.5	C.L.	32.35	17.81	14.44
15 - 30	3.6	29.7	39.3	27.4	C.L.	33.51	18.53	14.98
30 - 45	3.5	28.5	38.8	28.2	C.L.	32.52	17.96	14.56
45 - 60	3.8	28.7	39.6	27.9	C.L.	32.28	18.61	13.67

*Particle size distribution after [13] and moisture retention after [14]; C.L.: Clay Loam; F.C.: Field Capacity; W.P.: Wilting Point; A.W.: Available Water.

Table 2. Some chemical properties of the soil (saturated extracted)*.

Sample depth, cm	pH 1:2.5	EC dS/m	Soluble cations, meq/L				Soluble anions, meq/L			
			Ca^{2+}	Mg^{2+}	Na^{+}	K^{+}	CO_3^{2-}	HCO_3^{-}	SO_4^{2-}	Cl^{-}
0 - 15	7.3	0.35	1.5	0.39	1.52	0.12	0	0.31	1.52	1.67
15 - 30	7.2	0.36	1.51	0.44	1.48	0.14	0	0.41	1.56	1.63
30 - 45	7.3	0.34	1.46	0.41	1.4	0.13	0	0.39	1.41	1.63
45 - 60	7.4	0.73	2.67	1.46	3.04	0.12	0	0.67	2.86	3.82

*Chemical properties after [15].

Table 3. Some chemical properties of irrigation water.

pH	EC dS/m	Soluble cations, meq/L				Soluble anions, meq/l				SAR
		Ca^{2+}	Mg^{2+}	Na^{+}	K^{+}	CO_3^{2-}	HCO_3^{-}	SO_4^{2-}	Cl^{-}	
7.3	0.37	0.76	0.24	2.6	0.13	0	0.9	0.32	2.51	1.14

$$d = AMD \cdot ASW \cdot Rd \cdot P \quad (2)$$

where: AMD = allowable soil moisture depletion (%), ASW = available soil water, (mm water/m depth), Rd = effective root zone depth (m), or irrigation depth (m), and p = percentage of soil area wetted (%).

$$AW(v/v\%) = ASW(w/w\%) \cdot B.D. \tag{3}$$

where: B.D. = soil bulk density ($gm \cdot cm^{-3}$).

Irrigation Intervals used was 4 days under both closed circuits and traditional drip irrigation systems.

2.2. Measuring the Seasonal Evapotranspiration (ETc)

The (ETc) was computed using the Class Pan evaporation method for estimating (ETo) on daily basis was taken from nearest meteorological station as showing in **Table 4**.

The modified pan evaporation equation to be used:

$$ETo = KpEp \tag{4}$$

where: ETo = reference evapotranspiration [$mm \cdot day^{-1}$], Kp = pan coefficient of 0.76 for Class A pan placed in short green cropped and medium wind area. Ep = daily pan evaporation ($mm \cdot day^{-1}$), seasonal average is [7.5 $mm \cdot day^{-1}$] [16].

The reference evapotranspiration (ETo) is then multiplied by a crop coefficient Kc at particular growth stage to determine crop consumptive use at that particular stage of maize growth.

$$Etc = EToKc \tag{5}$$

The reduction factor (Kr) was calculated using Eq.6.

$$Kr = GC + 1/2(1 - GC) \tag{6}$$

where: GC = ground cover percentage.

Irrigation efficiency (Ea) was calculated by '

$$Ea = Ks \cdot Eu \tag{7}$$

where: Ea = irrigation efficiency (%), Eu = emission uniformity (%) and Ks = reduction factor of soil wetted.

The gross irrigation water requirements IWRg (mm depth) were calculated according to:

$$IWRg = IWRn \cdot Ea + Lr \quad (8)$$

where: IWRg = the gross irrigation water requirements, IWRn = the net irrigation water requirements and Lr = the extra amount of water needed for leaching.

Soybean seeds were (Glycine max-L, Rils-75) Varity was cultivated on April 15th. The distance between rows was 0.7 m and 0.15 m between plants in the row. Plants densities were 55,500 plants per fed according to (ISU). Each row was irrigated by a single straight lateral line in the closed circuits and traditional drip irrigation plots. **Figure 1** shows that the total experimental area was 4536 m^2. Under each of the tested drip irrigation circuits, plot areas of Lateral lines lengths were 168, 252 and 336 m^2 under LLL1, LLL2 and LLL3, respectively. Soybean was harvested on September 11. Irrigation season of Soybean was ended 10 days before harvest. Fertilization program had been done according to the recommended doses throughout the growing season using fertigation technique. These amounts of fertilizers NPK (20-20-10), were 74.6 kg/fed of N and 33.0 kg/fed of K_2O. While 60.5 kg/fed of P_2O_5. For all plots, weed and pest control applications followed recommendations of Soybean yield.

2.3. Measurements of Soybean Plant Growth Yield and Quality

Plant measurements and observations were started 21 days after planting, and were terminated on the harvest date. Yield and components including leaf area (cm^2) by Plano meter, plant height

Table 4. Water requirements for soybean grown at experimental site.

Month	Apr.	May	Jun.	Jul.	Aug.	Sep.
Epan (mm/day)	6.34	6.92	7.97	9.59	9.32	7.17
Kp			0.76			
Kc	0.72	0.82	0.93	1.18	1.2	1.23
Kr	0.25	0.63	0.95	1	1	1
ETo (mm/day)	4.75	5.26	6.06	7.29	7.09	5.45
ETc (mm/day)	0.85	2.72	5.35	8.6	8.51	6.7
Ks			100% (1.00)			
Eu			90% (1.11)			
Lr			10%			
Growth stage	Planting (Establishment)	Rapid vegetative growth	Flowering-seed fill		Maturation and harvesting	
Length of growth stage	15Apr.-14May	14May-13 Jun.	13Jun.-12 Aug.		12Aug.-11Sep.	
Number of days (Irri. season)	30	30	61		31	
IRn (mm/month)	15.0	92.8	176.6	293.3	290.2	81.1
IRg (mm/month)	49.3	158.8	198.6	264.5	268.2	27.3

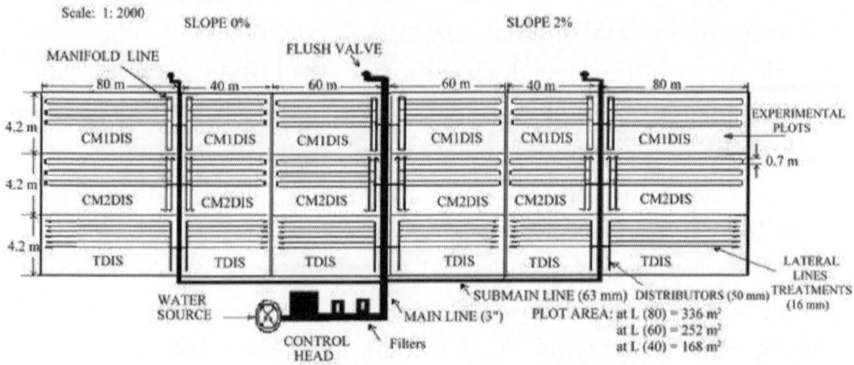

Figure 1. Layout of the field experimental plots: using DIC, (CM2DIS, CM1DIS and TDIS); and (LLL_1 = 40 m; LLL_2 = 60 m and LLL_3 = 80 m) treatments.

(cm) by meter measuring, total grain and straw (Kg/fed) by balance; oil and protein (g/kg). Oil and protein determined in Grain Quality Laboratory using near-infrared analysis. All plant samples were dried at 65°C until constant weight. Grain yield was determined by hand harvesting the 8 m sections of three adjacent center rows in each plot and was adjusted to 15.5% water content. In all treatments plots, the grain yields of individual rows were determined in order to evaluate the yield uniformity among the rows.

MSTATC program (Michigan State University) was used to carry out statistical analysis. Treatments mean were compared using the technique of analysis of variance (ANOVA) and the least significant difference (L.S.D.) between systems at 1% [17].

3. RESULTS AND DISCUSSION

3.1. Leaf Area (cm^2) and Plant Height (cm)

Table 5 indicated the effects of DIC and LLL treatments on leaf area and plant height, treatments used could be ranked in following descending orders: CM2DIS ≥ CM1DIS < TDIS and LLL1 < LLL2 < LLL3. Differences in leaf area and plant height between means of the two factors studied were significant at the 1% level except that between CM2DIS and CM1DIS for both leaf area and plant height and between LLL1 and LLL2 for plant height. The effects of DIC × LLL on plant height and leaf area were significant at the 1% level. The superiority of the studied growth parameters under (CM2DIS; CM1DIS relative to TDIS) and (LLL1; LLL2 relative to LLL3) can be noticed. This superiority was due to improving both water and fertilizers distribution uniformity [8-12].

3.2. Grain and Straw Yield (kg/fed)

Table 5 showed the effect of DIC and LLL treatments on grain and straw yield (kg/fed). They could be ranked in following descending order: CM2DIS > CM1DIS > TDIS and LLL1 > LLL2 > LLL3. Differences in grain and straw yield between means of any two treatments were significant at the 1% level except that between CM1DIS and CM2DIS in straw yield. The effects of the DIC × LLL on both grain and straw yield were significant at the 1% level. The highest and lowest values of both grain and straw yield were recorded in the interacttions: CM2DIS × LLL1 and TDIS × LLL3, respectively. This superiority was due to improving both vegetative growth, water and fertilizers distribution uniformity [8-13].

3.3. Oil and Protein (g/kg)

Data in **Table 5** indicated the effects of DIC and LLL treatments on Soybean oil and protein production (g/kg), both of them could be ranked in the following ascending orders: TDIS < CM1DIS < CM2DIS and LLL3 < LLL2 < LLL1, respectively. In respect to the main effect of (DIC) on both of oil and protein, one can notice that, the differences in both of oil and protein were significant among all DIC and LLL treatments at the 1% level except that between CM1DIS and CM2DIS in oil. The highest and lowest oil and protein were obtained in the interactions CM2DIS × LLL1 and TDIS × LLL3, respec tively.

Table 5. Effect of irrigation circuits designs and lateral lines lengths on soybean plants growth yield and quality (operating pressure = 1 atm and slope = 0%).

DIC	L.L.L. (m)	Growth, yield and quality characteristics (average)						Water use efficiency (kg/m^3)	
		Leaf area (cm^2)	Plant height (cm)	Yield (kg/fed)		Quality (g/kg)			
				Grain	Straw	Oil	Protein	WUEg	WUEs
	40	7.91a	94.38a	657.5a	588.6a	183.5a	366.2a	0.150a	0.134a
CM2DIS	60	7.62cb	92.26dc	648.3cb	557.5d	181.2d	364.1d	0.148b	0.127d
	80	6.56f	91.15h	641.6db	531.8h	178.1g	361.8g	0.146d	0.121h
	40	7.64b	94.29ba	642.2b	583.7b	183.3b	365.3b	0.146cd	0.133b
CM1DIS	60	7.43dc	92.35e	628.3e	546.4e	180.6e	363.6e	0.143e	0.125e
	80	6.22h	91.52f	597.7g	537.2g	177.6h	361.3h	0.136g	0.123g
	40	6.85e	92.11e	605.3f	574.3c	182.4e	365.1c	0.138f	0.131e
TDIS	60	6.51g	91.18g	593.4hg	542.8fe	179.7f	362.2f	0.135h	0.124f
	80	5.92i	90.23ih	586.2i	519.6i	176.3i	360.4i	0.134ih	0.119i
(1) × (2)	$LSD_{0.01}$	0.23	0.12	8.2	4.8	0.14	1.3	0.002	0.001
	CM2DIS	7.36a	92.59a	649.1a	559.3a	180.9a	364.0a	0.148a	0.128a
(1) Means	CM1DIS	7.09ba	92.72ba	622.7b	555.8ba	180.5ba	363.4ba	0.142b	0.127ba
	TDIS	6.43c	91.17c	595.0c	545.6c	179.5c	362.6c	0.136c	0.124c
	$LSD_{0.01}$	0.48	0.51	15.8	7.2	0.6	0.7	0.004	0.002
	40	7.47a	93.60a	635.0a	582.2a	183.1a	365.5a	0.145a	0.133a
(2) Means	60	7.19ba	91.93b	623.3b	548.9b	180.5ba	363.3b	0.142b	0.125b
	80	6.23c	90.97c	608.5c	529.5c	177.3c	361.2c	0.139c	0.121c
	$LSD_{0.01}$	0.72	0.71	12.4	26.4	3.1	1.9	0.003	0.005

DIC: Irrigation circuit design; L.L.L.: Lateral line length; CM2DIS: Closed circuits with tow manifolds separately; CM1DIS: Closed circuits with one manifold; TDIS: Traditional drip irrigation system.

3.4. WUEg and WUEs (kg/m^3)

Table 5 showed the effects of both DIC and LLL treatments on WUEg and WUEs (kg/m^3). We can notice that the changes in WUEg and WUEs took the same trend of growth, grain, straw, parameters and thus took the trend of quality (oil; protein). Concerning the positive effects of DIC and LLL treatments on WUEg and WUEs, they could be ranked in following descending orders: CM2DIS > CM1DIS > TDIS and LLL1 > LLL2 > LLL3, respecttively. Differences in WUEg and WUEs between means of any two treatments were significant at the 1% level except that between CM1DIS and CM2DIS in WUEs. The effect of the DIC × LLL on WUEg and WUEs were significant at 1% level. The highest and lowest values of WUEs were obtained at CM2DIS × LLL1 and TDIS × LLL3, respectively. We can notice that the Soybean WUEg, WUEs oil and protein took the same trend of other vegetative growth and yield parameters, and this finding could be attributed to the close correlation between vegetative growth; grain yield from side and quality of oil and protein production from the other one [8- 12].

4. CONCLUSIONS

Data on hand could be stated in the following:

1) According to the mean values of soybean crop growth (leaf area; plant height), yield (grain and straw), both oil and protein content and water use efficiency, the treatment used could be ranked in the following ascending orders: TDIS < CM1DIS < CM2DIS and LLL3 < LLL2 < LLL1;

2) Differences in the means of the studied data among treatments used were significant at the 1% level;

3) The effects of the DIC × LLL on the data obtained were significant at the 1% level;

4) The highest values of the obtained data and the lowest ones were achieved in the following interactions: CM2DIS × LLL1; TDIS × LLL3, respectively.

REFERENCES

1. Fao (1979) Yield response to water. FAO Irrigation and Drainage Paper, 33, 193.

2. Harrison, K.A. (2009) Irrigation survey. Cooperative Extension Service, College of Agricultural and Environmental Sciences, The University of Georgia, Tifton.http://www.nespal.org/sirp/agwateruse/facts/survey/.%5C2009.0122.IrrSurvey08%20misc%20pub.pdf

3. United States Department of Agriculture—National Agriculture Statistics Service, 2009.http://www.nass.usda.gov/Quick_Stats/

4. Dogan, E., Kirnak, H. and Copur, O. (2007) Deficit irrigations during soybean reproductive stages and CROPGRO-soybean simulations under semi-arid climatic conditions. Field Crops Research, 103, 154-159.

5. Sincik, M., Candogan, B.N., Demirtas, C., Büyükacangaz, H., Yazgan, S. and Goksoy, A.T. (2008) Deficit irrigation of soybean [Glycine max (L.) Merr.] in a sub-humid climate. Journal of Agronmy and Crop Science, 194, 200- 205.

6. Bajaj, S., Chen, P., Longer, D.E., Shi, A., Hou, A., Ishibashi, T. and Brye, K.R. (2008) Irrigation and planting date effects on seed yield and agronomic traits of earlymaturing soybean. Journal of Crop Improvement, 22, 47- 65.

7. Gercek, S., Boydak, E., Okant, M. and Dikilitas, M. (2009) Water pillow irrigation compared to furrow irrigation for soybean production in a semi-arid area. Agricultural Water Management, 96, 87-92.

8. Mansour, H.A. (2012) Design considerations for closed circuits of drip irrigation system. Ph.D. Thesis, Ain Shams University, Cairo.

9. Tayel, M.Y., Mansour, H.A. and Lightfoot, D. A. (2012) Effect of different closed circuits and lateral line lengths on pressure head and friction loss. Agriculture Science Journal, 3, 392-399.

10. Tayel, M.Y., Mansour, H.A. and Lightfoot, D. A. (2012) Effect of different closed circuits and lateral line lengths on flow velocity and velocity head. Agriculture Science Journal, 3, 531-537.

11. Tayel, M.Y., Mansour, H.A. and El-Gindy, A.M. (2012) Effect of different closed circuits and lateral line lengths on dripper and lateral discharge. Journal of Applied Sciences Research, 8, 2725-2731.

12. Tayel, M.Y., Mansour, H.A. and El-Gindy, A.M. (2012) Effect of different closed circuits and lateral line lengths on uniformity coefficient and coefficient of variation. Journal of Applied Sciences Research, 8, 2741-2748.

13. Gee, G.W. and Bauder, J.W. (1986) Particle-size analysis. In: Klute, I.I., Ed., Methods of Soil Analysis, Soil Science Society of America, Madison, 383-412.

14. Klute, A. (1986) Moisture retention. In: Klute, A. Ed., Methods of Soil Analysis, Soil Science Society of America, Madison, 635-662.

15. Rebecca, B. (2004) Soil survey laboratory methods manual. Soil Survey Laboratory Investigations Report, No. 42.

16. Allen, R.G., Pereira, L.A., Raes, D. and Smith, M. (1998) Crop evapotranspiration. FAO Irrigation and Drainage Paper, 56, 293.

17. Steel, R.G.D. and Torrie, J.H. (1980) Principles and procedures of statistics. A biometrical approach. 2nd Edition, McGraw-Hill Book Company, Tokyo.

8

THE OPTIMIZATION OF IRRIGATION NETWORKS USING GENETIC ALGORITHMS

Francisco Fernando Noronha Marcuzzo[1], Edson Cezar Wendland[2]

[1]Department of Hydrology Applied, Geological Survey of Brazil, Porto Alegre, Brazil

[2]Department of Hydraulic, School of Engineering of São Carlos, University of São Paulo, São Carlos, Brazil

ABSTRACT

Mathematical-computational optimisation models of irrigation networks with a distributed flow that are capable of providing hydraulic data are important for understanding the behaviour of a system in relation to the distribution of the hydraulic head (energy) and the pressure in the pipes of the network. The objective of this study was to examine the distribution of the parameters of hydraulic irrigation pipes, which were optimised using genetic algorithms. The degree of the optimisation was evaluated with the help of the genetic algorithms based on the diameters of stretch of the network: two for the lateral lines, four for the derivation lines, four for the secondary lines and one for the main line. A MatLab code was developed that considered all of the losses of energy, both distributed losses and those at specific locations between the beginning of the network and the pump system. The sensitivity analysis was based on the variations in the slope of the ground (0%, 2.5% and 5%). The results show that for pipes with a distributed flow, the influence of the behaviour of the kinetic energy in the pipe contributed to the distance between the energy lines and the piezometric lines at the beginning of each stretch after the decrease in the diameter of the pipes. At the end of the pipes, the values of the energy lines and the piezometric lines were very similar, and they were essentially the same for the final emitter.

Keywords: Conduits under Pressure, Hydraulic Head, Micro-Sprinkler Irrigation, Mathematical Optimisation

1. INTRODUCTION

Localised irrigation is the application of small amounts of water, normally the volume that the plant will use for a few days, with an application frequency greater than in conventional irrigation. These systems normally use small amounts of energy for pumping, pressure and flow. However, localised irrigation networks may display significant variations in the drippers, which leads to a lack of uniformity of their implementation in the field.

In a study of the hydraulic performance of the lateral lines of drip tubes, [1] has stated that localised irrigation allows for the best possible control over the irrigation water applied because it operates with reduced flows and higher irrigation frequencies. In their field research on the hydraulic parameters of micro-sprinkler systems, [2] has observed the solid functioning of the system, with high coefficients of uniformity, good efficiency of application and little variation in the flow and pressure of the system, indicating excellent quality of the implementation of the irrigation system. [3] has observed that the hydraulic performance of a low-pressure irrigation system that was designed and fit dimensionally using a computer program and installed in the field was considered satisfactory because it exhibited good efficiency and excellent hydraulic uniformity.

In a study of the optimisation of the network with different slopes, [4] has reported that the use of two diameters in the derivation line improved the uniformity of the flow of the derivation line. In addition, [5] [6] have performed optimisation studies and analyses on the dimensions of the irrigation networks by considering the different options for the operation of the system as the basis for the decision-making process. These studies showed excellent results for mathematical optimisation, which is an important decision-making tool in the compromise between the best system management and the required dimensions of the distribution network.

In a study using different indices to define the standard water use in irrigation systems, farmers cultivating onions using a system with micro-sprinklers have increased their production, while using less water [7] . In addition, a study comparing two irrigation methods (dripping and sprinklers) for sunflower seeds plants in the Mediterranean has shown that the water volume is important for increased productivity and the yield of sunflower oil [8] .

The local head loss, which occurs when the flow is disrupted by an obstruction in the tube, is a relevant criterion in the evaluation of lateral drip lines, especially when self-compensating drippers are used [9] . The high uniformity of the distribution of water allows for the optimisation of the use of water, energy and nutrients; to guarantee this, appropriate hydraulic design is indispensible, requiring that the hydraulic characteristics of the components are known [10] .

Considering the optimisation objectives for the rehabilitation of a water supply network, the use of a standard genetic algorithm improves both the convergence speed and the rationality and viability of the solutions for the problems of mathematical programming [11] .

[12] has conducted a study of the irrigation management in the Ceyhan River Basin in Turkey using dynamic programming with successive approximations (DPSA). Given that the benefit may be defined as an ideal value when considering the balance between irrigation water and energy production, it was observed in the above study that when the irrigation demand was increased, the production of energy and the benefits of producing energy decreased.

In a dam project based on the simulation of optimisation through a genetic algorithm, [13] has demonstrated a significant cost reduction using a Non-linear Optimisation Formulation (NLOF).

This article presents a hydraulic study of a water distribution network for localised irrigation that was obtained through optimised simulation using genetic algorithms, including the consideration of the variation in the slope of the ground. Thus, we performed a numeric simulation with the optimised variables to analyse the behaviour of the hydraulic design elements throughout the pipes where there were derivations of the flow distributed in the tubing.

2. MATERIAL AND METHODS

The distribution network for localised irrigation was installed in an area that included 68.4 useful hectares. Figure 1 presents the distribution and the dimensions of the hydraulic network.

The hydraulic network of the localised irrigation system studied in this work consisted of emitters (micro-sprinkles or drippers), lateral lines (in which the emitters were located), derivation lines (which distributed water to the lateral lines), secondary lines (which distributed water to the derivation lines), the main line or the adductor line (which distributed water to the secondary lines), suction tubing, valves, registers, filters (sand and disc) and the control panel.

Table 1 presents in detail the different stretches of piping used in this study, their lengths, whether the ground was even or upward or downward sloping (based on the direction of the flow of water), the material used in the production of these elements and the internal diameters used in each stretch of optimised tubing. The slopes of the ground were 0%, 2.5% and 5%.

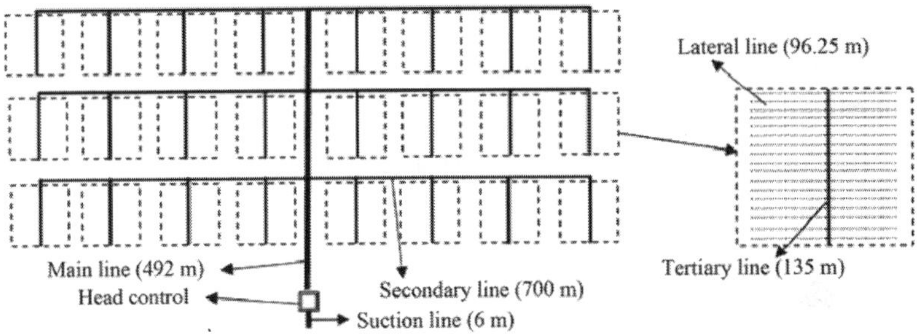

Figure 1. The design of the micro-sprinkler network with the respective lengths of pipe used.

Table 1. Details of the adopted irrigation network.

Stretch of the Network	Position of the Tubing	Material	Length	Internal Diameter mm
Lateral Line	Level	Polyethylene	96.25	21.7; 17.8
Derivation Line	Decline	PVC	135	120.0; 97.6; 72.5; 48.1
Secondary Line	Level	PVC	700	204.2; 144.0; 120.0; 97.6
Main Line	Incline	PVC	492	204.2
Suction Line	Incline or Level	PVC	6	262.0

To calculate the adjustment factor for the decrease in the loss of pressure due to friction of the tubing, the following adjustment equation proposed by [14] was used:

$$FLi = \frac{NTDi \times \left(\left(\frac{1}{m+1} \right) + \left(\frac{1}{2NTDi} \right) + \left(\frac{(m-1)^{0.5}}{6NTDi^2} \right) \right) + xLi - 1}{NTDi + xLi - 1} \quad (1)$$

in which:

FLi is the adjustment factor for a decrease in the loss of pressure due to friction of the tubing i;

$NTDi$ is the total number of derivations for the tubing i ;

m is the exponent for the equation of unitary head loss; and

xLi is the ratio of the distance from the beginning of line i to the first derivation, in relation to the distance between the other derivations.

To calculate the coefficient of friction, the following general equation proposed by [15] was used, which is used for rolling, smooth turbulent, transition and rigorous turbulent flows:

$$fin = \left\{ \left(\frac{64}{Rein} \right)^8 + 9.5 \left[\ln \left(\frac{\varepsilon}{3,7 Din} + \frac{5.74}{Rein^{0,9}} \right) - \left(\frac{2500}{Rein} \right)^6 \right]^{-16} \right\}^{0,125} \quad (2)$$

in which:

fin is the coefficient of friction for the calculation of the head loss in the tubing in line i in stretch n (dimensionless);

$Rein$ is the Reynolds number in line i in stretch n (dimensionless);

ε is the absolute roughness of the tubing used in line i (mm); and

Din is the diameter of the pipes in line i in stretch n (m).

The unit head loss was determined using the Darcy-Weisbach equation [16] , as follows:

$$Jin = \frac{fin}{Din} \frac{Vin^2}{2g} \quad (3)$$

in which:

Jin is the unit head loss in line i in stretch n ($m \cdot m^{-1}$);

fin is the coefficient of friction in line i in stretch n (dimensionless);

Din is the diameter of the pipes in line i in stretch n (m);

Vin is the velocity of the water in line i in stretch n ($m{\cdot}s^{-1}$); and

g is the acceleration of gravity ($m[s^2]^1$).

The sum of the unit head loss with a distributed slope of the ground was calculated using the following:

$$fin = (Jin\ FLi\ LLin) + PCLin + DZLLin \tag{4}$$

in which:

$hfin$ is the sum of the head loss and the slope of the ground in line i in stretch n (mca);

Jin is the unit head loss in line i in stretch n ($m{\cdot}m^{-1}$);

FLi is the adjustment factor for the decrease in the pressure loss due to friction in the tubing i;

$LLin$ is the length of line i in stretch n (m);

$PCLin$ is the local head loss from the emitters in the tubing in line i in stretch n (m); and

$DZLLin$ is the incline or decline distributed by the loss of water in line i in stretch n (m).

The optimisation process of the dimensions began with the lateral line, in which the emitters were located. The flow was determined for each emitter according to the water needs of the plants that would be irrigated. The sum of the flow of all of the emitters indicated the flow that should be available at the beginning of the lateral line. The calculation of the dimensions was performed stretch by stretch, from the beginning of the line until the end, subtracting the distributed flow from the previous stretch. This principle was adopted for the lateral lines, the derivation lines and the secondary lines. For the main line (adductor) and the suction line, the calculation of the dimensions was not performed stretch by stretch because there were no derivations that simultaneously received water.

Because the dimensioning should be performed stretch-by-stretch in a network with a distributed flow, pre- measurement of the tubes in the network was necessary.

The determination of the flows in the tubes of the network is fundamentally important in the design of the code for the capability function, whether it be the entire flow at the beginning of the piping or the fractioned flow for each derivation (emitters and piping).

The methodology that was used for the economic analysis implemented by the computational code is described below. In the calculation of the total cost, the fixed cost of installation and the variable operational costs were both considered.

Using annual interest, the capital recovery factor (CRF) provided a coefficient that allowed for the calculation of the fixed annual cost of an investment based on the value of the investment. This calculation took into account the useful life of the equipment and the per annum interest rate using the following equation:

$$CRF = \frac{J(J+1)^V}{(J+1)^V - 1} \quad (5)$$

in which:

CRF is the capital recovery factor (decimal);

J is the annual interest rate (decimal); and

V is the useful life of the equipment (years).

The calculation of the fixed annual cost took into account the CRF and the price of acquiring the system, using the following equation:

$$CF = FRC \times CEQ \quad (6)$$

in which:

CF is the fixed annual cost (R$·year^{-1}·ha^{-1});

FRC is the capital recovery factor (decimal); and

CEQ is the price of acquiring the irrigation system (R$·ha^{-1}).

The costs of both energy and water were included in the calculation of the annual variable costs of irrigation.

For the cost of energy, it is important to consider the method of charging for electricity. According to a resolution by the National Agency of Electrical Energy DOU no. 313 from 06/04/2006 [17] , this irrigation project fell under Group B-the rural sector category-with a payment for consumption of only 0.1768 R$·kW^{-1}.

The electricity provider for the state of Sao Paulo (Power Company of São Paulo), designates a discount of 60% on the tariff for the rural sector between the hours of 21:30 and 06:00 (a "window" of use of 8.5 hours), which would result in a rate of 0.0707 R$·kW^{-1} for this time frame. Through heuristics, the system could be adjusted to function for 4.25 hours in two cycles per day, reducing the flow through the project and generating savings in the piping and power required; another option would be to use a cycle of 8.5 hours, depending on the preferred objective. This adjustment could be completed by increasing the flow of the emitters according to the flow curve versus the pressure provided by the manufacturer or by simply selecting another emitter. The choice of adjustment would depend on how many days the irrigation would last and how many sectors would be irrigated per day, with the operational units based on the system that would be installed.

The fee for the use of water was calculated based on the evapotranspiration of the crop, the number of irrigation days per month of operation of the system per year and the price of water (R$·m^{-3}). To make the model broader and more realistic, a water cost of 0.01 R$·m^{-3} was adopted as an entrance parameter for the operation cost.

The total cost (the capability function to be optimised) was the sum of the fixed annual costs (installation of the irrigation system) and the variable annual costs (operational costs), calculated using the following equation:

$$f = \frac{(CEQ \cdot FRC) + CTE + CAGT}{AUI} \tag{7}$$

in which:

f is the total cost of the irrigation network (R$·year^{-1}·ha^{-1});

CTE is the total cost of electricity (R$·year^{-1});

$CATG$ is the total cost of water (R$·year^{-1}); and

AUI is the useful irrigated area (ha).

The code for the optimised design was developed using MatLab. For the optimisation, the genetic operators were used from the MatLab toolbox.

The sensitivity analysis evaluated the effects of the operational unit of the system over the hydraulics (energy and pressure), based on the variation in the slope of the terrain. Sensitivity analysis is the measurement of the magnitude of the alteration of the final result caused by a predetermined change of one or more factors of the project, which means that the projected model is tested to determine the most relevant elements for the design decision. Thus, with various design parameters, it would be possible to select those parameters that are most sensitive, that is, those that have the most the influence over the analysis [18] [19] .

The complete algorithm for the optimized design is in [20] and each analysis step for better presentations is in [21] [22] . The sensitivity analysis can be finding in [23] .

3. RESULTS AND DISCUSSION

The Hydraulic Analysis of the Lateral Line

Figure 2 presents the variation in the velocity and the distributed flow in the lateral line in relation to the internal diameter of the piping.

In Figure 2, the direct influence of the diameter of the tubing is apparent, with a distributed flow and with an increase in the velocity of the flow when the diameter of the first stretch of tubing experienced a contraction down to the diameter of the second stretch of tubing. This increase in the flow velocity occurred when there was a swift decrease in the internal area of the tubing and the flow in the closest emitters was insufficient to decrease the velocity around the point of contraction. After the 13th emitter, 5 emitters were necessary (after the 19th emitter), such that that velocity of the flow was once again less than that of the 13th emitter in the lateral line.

Figure 3 shows the Reynolds number for the emitters and the regions that represent turbulent, transitional and rolling flows, as determined according to Harpa de Nikuradse [16] .

The descriptive analysis in Figure 3 shows that the flow of the lateral line began and remained turbulent in the tubing, with 25 emitters in this part of the flow, until passing through a short stretch of flow in transition and then ending in a rolling flow. With this outflow and with a distributed flow, the velocity and the diameter of the tubing (as calculated in Figure 2) had a direct result on the variation in the Reynolds number and the type of flow in the lateral line. This effect on the type of flow demonstrated that the selection of the equation to calculate the friction factor was particularly important for smooth tubes, such as those made of polyethylene or PVC (often used in the design of irrigation networks), because the type of flow may change, as in Figure 3, in which the polyethylene tubing was under a distributed flow.

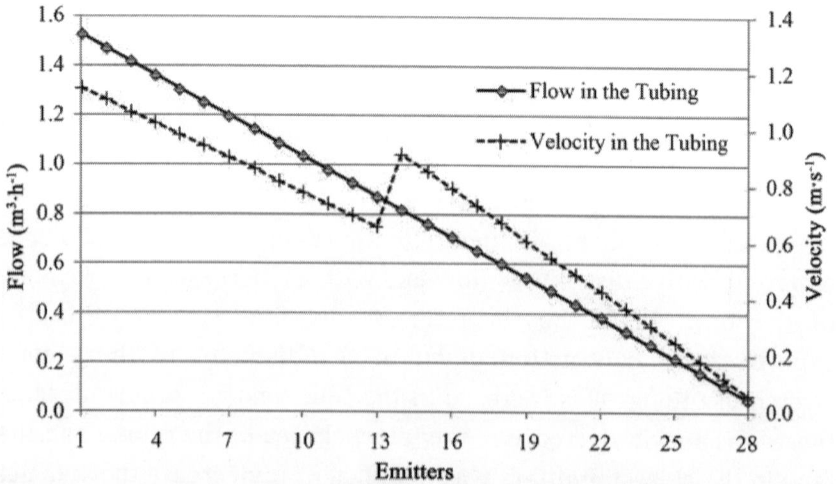

Figure 2. The distributed flow and velocity calculated in the design of the lateral line, with two different internal diameters in the tubing.

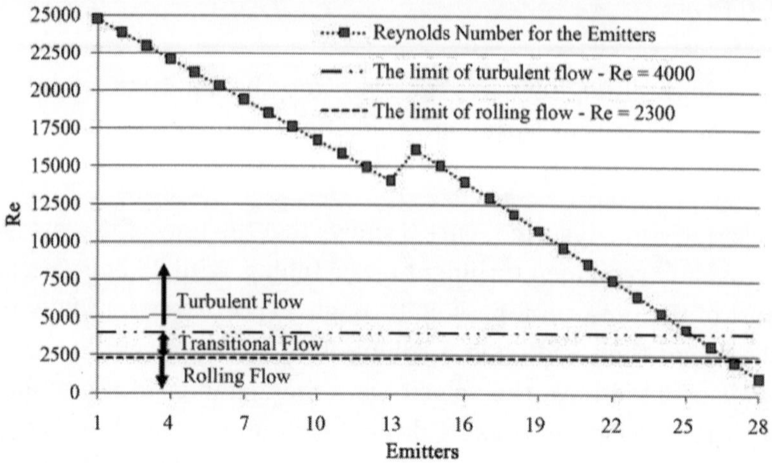

Figure 3. The Reynolds number for the emitters and the regions that represent rigorous, transitional and rolling flows.

Figure 4 presents the local head loss in the contraction of the tubing and the variation in the local head loss throughout the emitters and the head loss unit between the emitters, the friction factor and the kinetic energy that all resulted from the scaling of the lateral line with a distributed flow. With a distributed flow, the variation in these scaling parameters corresponded to the tendency determined by the internal diameter of the tubing and the velocity of the outflow (Figure 2).

The local head loss with the contraction of the tubing was small (0.0045 mca) in comparison with the head loss unit (0.0650) and the head loss between the emitters attached to the lateral line (0.0212 mca) at the point of contraction of the tubing. The variation in the friction factor (from 0.0246 mca in the first emitter up to 0.0594 mca in the last emitter) throughout the lateral line tended to be inversely proportional to the variation in the Reynolds number presented in Figure 3. However, it can be seen that at emitter point 26, there was a slight tendency for a decrease in the friction factor and that at emitter point 27, there was an exacerbated decrease followed by a swift and accentuated increase at the final emitter point. This phenomenon occurred because of the intrinsic characteristics of the calculation of the friction factor and its variations and may be observed by analysing the polyethylene tubing (0.0015) in the Moody diagram.

The influence of the kinetic energy (Figure 4) in the tubing contributed to the distance between the energy lines and the piezometric line at the beginning of the first and second stretches of the tubing. At the end of the tubing, after the 26th emitter, the values remained very similar, and they were essentially equal for the final emitter (28), where the kinetic energy was very close to zero.

The small amount of variation (9.1%) between the highest (16.66 mca) and the lowest value of the piezometric head (15.14 mca) at the emission points in the lateral line caused a greater uniformity of emission in the operational unit.

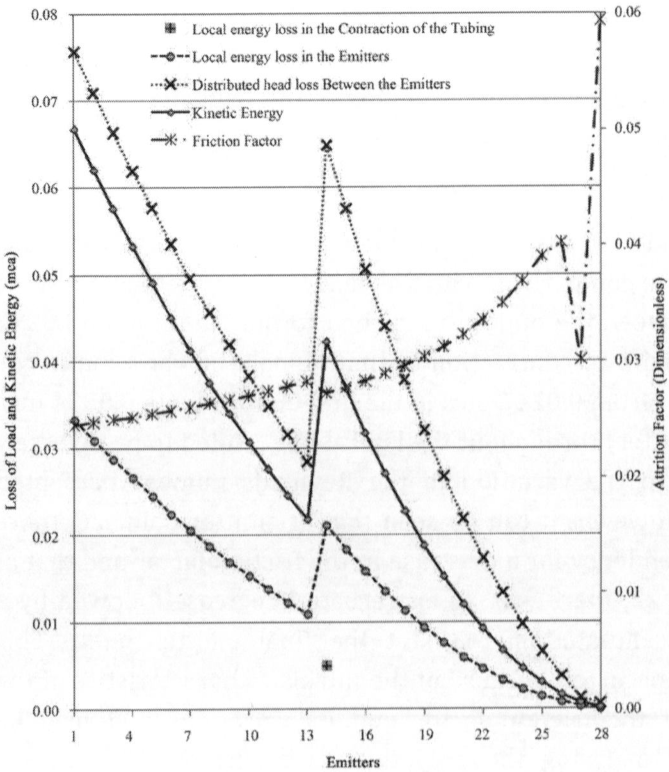

Figure 4. The local head loss in the contraction of the tubing and the variation in the local head loss throughout the emitters and the head loss unit between the emitters, the friction factor and the kinetic energy in the lateral line under a distributed flow. The data were obtained from the design of the network.

The variation in the flow throughout the lateral line tended to stabilise concomitant with the stabilisation of the piezometric line, whereas the flow in each emitter was determined as a function of the pressure head at the emission point. More details on the influence of the variation of the piezometric line in the floss of the emitters will be presented in the hydraulic analysis of the entire operational unit (lateral lines attached to a derivation line) with the different slopes in the terrain.

Figure 5 presents the relationship between the energy lines and the piezometric lines obtained from the network plan and the flow in the emitters, with the piezometric head in each emitter.

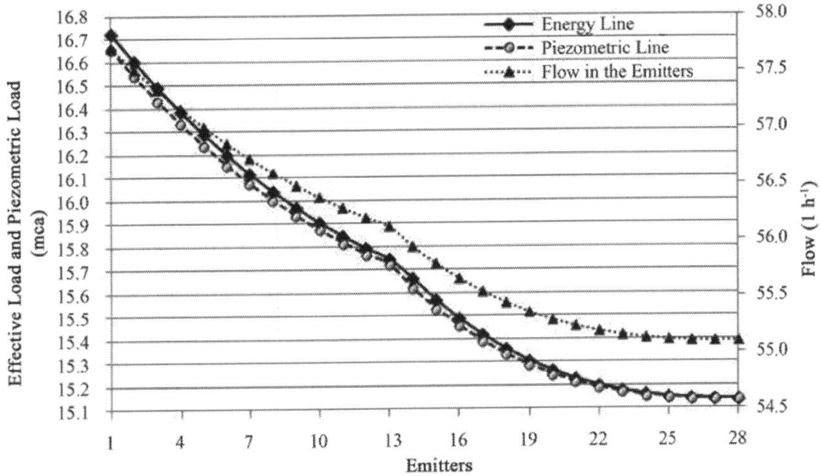

Figure 5. The variation between the energy lines and piezometric lines obtained from the dimensions of the network and the flow in the emitters, with the piezometric head in each emitter.

Figures 6-8 present the variation in the velocity of the outflow in the derivation line in comparison with the flow and the internal diameter of the tubing for the slopes of 0%, 2.5% and 5%, respectively, calculated using the dimensions of the irrigation network.

The velocity increased at the outflow, even when the tubing had a distributed flow, and thus, there was a tendency toward a decrease in the velocity at the outflow, which led to an accentuated decrease in the internal area of the tubing.

In Figures 6-8, a lower average velocity (1.49 $m \cdot s^{-1}$) is observed in the tubing for a 0% slope in the terrain compared with the tubing for the 2.5% (1.71 $m \cdot s^{-1}$) and 5% (1.80 $m \cdot s^{-1}$) slopes. Thus, there tended to be an increase in the average velocity in the tubing of the derivation line when there was an increase in the slope with the same distributed flow.

A greater magnitude change (1.59 $m \cdot s^{-1}$) was also noted between the minimum velocity (0.47 $m \cdot s^{-1}$) and the maximum velocity (2.05 $m \cdot s^{-1}$) for a slope of 0%, as compared with the magnitude change for a slope of 2.5% (2.00 $m \cdot s^{-1}$, with a maximum of 2.46$\cdot$m s^{-1} and a minimum of 0.47 $m \cdot s^{-1}$) and 5% (2.02 $m \cdot s^{-1}$, with a maximum of 2.49 $m \cdot s^{-1}$ and a minimum of 0.47 $m \cdot s^{-1}$).

Figures 9-11 present the variation in the Reynolds number along the derivation line for the slopes of 0%, 2.5% and 5%, respectively. The variation in the Reynolds number along the derivation line followed the tendency represented in Figures 6-8. With the contraction of the diameters, there was an increase in the outflow velocity, causing a local increase in the Reynolds number, which was directly reflected in the other hydraulic design parameters.

The smallest variation in the Reynolds number was obtained with a 0% slope in the terrain, with a value of 89.2%, followed by a slope of 2.5%, with 90.3% variation and, lastly, a slope of 5%, with a variation of 90.7%.

Figures 12-14 present the data obtained from the dimensioning of the irrigation network, with a local head loss with 3 contractions in the tubing, including the local head loss behaviour with the attachment of the lateral lines, the head loss unit between the lateral lines and the friction factor and kinetic energy of the derivation line under a distributed flow. The internal area of the tubing and the velocity of the outflow under a distributed flow determined the behaviour of these parameters in the dimensions of the tubing (Figures 6-8). The local head loss in the contraction of the tubing was the smallest at the first contraction (0.0247, 0.0395 and 0.0447 mca for 0%, 2.5% and 5% slopes, respectively) in comparison with the rest of the contractions. The local head loss estimates for the second contraction (0.0464, 0.0664 and 0.0628 mca for 0%, 2.5% and 5% slopes, respectively) were, in general, lower than those for the third contraction (0.0668, 0.0576 and 0.0873 mca for 0%, 2.5% and 5%

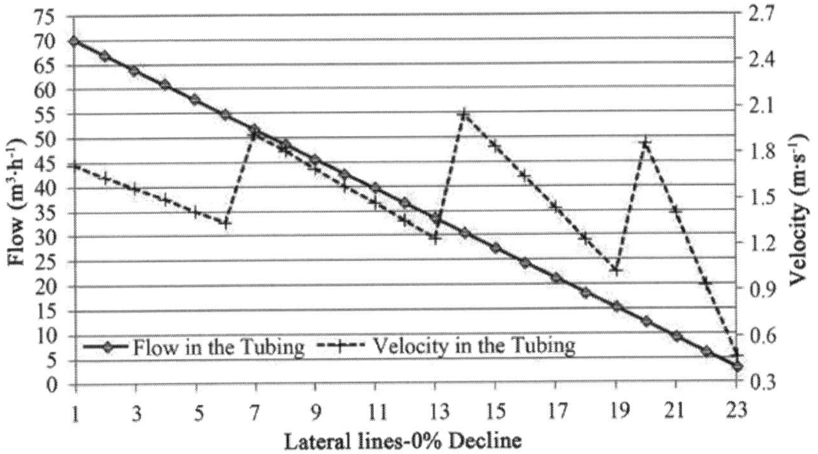

Figure 6. The variation in the calculated velocity in the dimensions of the derivation line, with a distributed flow in the tubing and different diameters in each stretch of the tubing (DLD1 = 120.0 mm, DLD2 = 97.6 mm, DLD3 = 72.5 mm and DLD4 = 48.1 mm) for a slope of 0%.

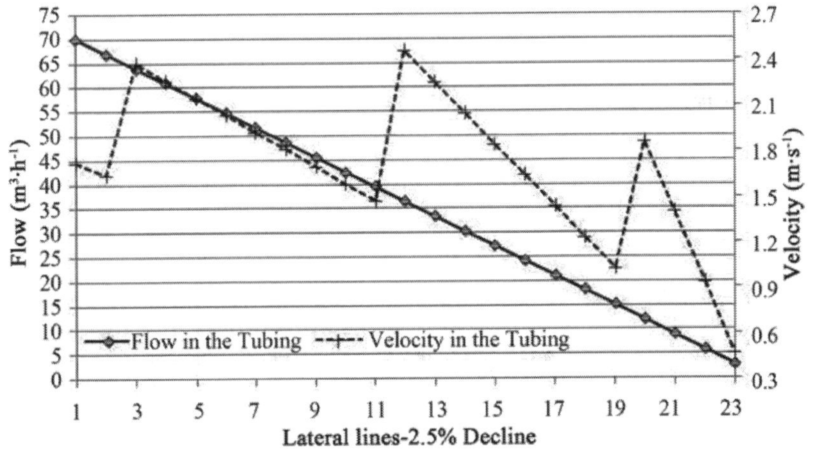

Figure 7. The variation in the calculated velocity in the dimensions of the derivation line, with a distributed flow in the tubing and different diameters in each stretch of the tubing (DLD1 = 120.0 mm, DLD2 = 97.6 mm, DLD3 = 72.5 mm and DLD4 = 48.1 mm) for a slope of 2.5%.

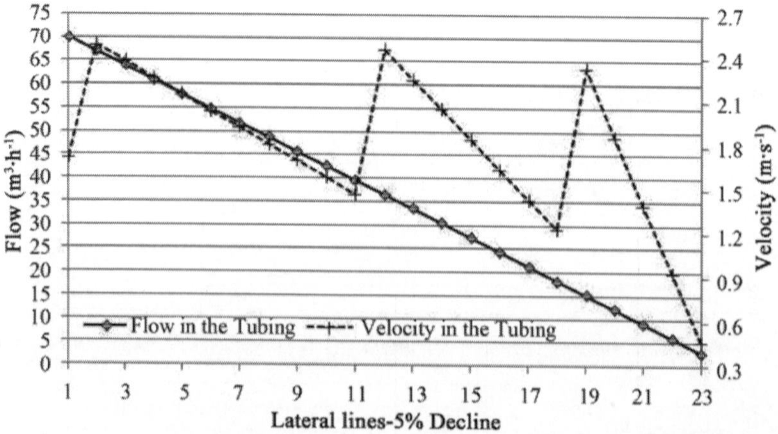

Figure 8. The variation in the calculated velocity in the dimensions of the derivation line, with a distributed flow in the tubing and different diameters in each stretch of tubing (DLD1 = 120.0 mm, DLD2 = 97.6 mm, DLD3 = 72.5 mm and DLD4 = 48.1 mm) for a slope of 5%.

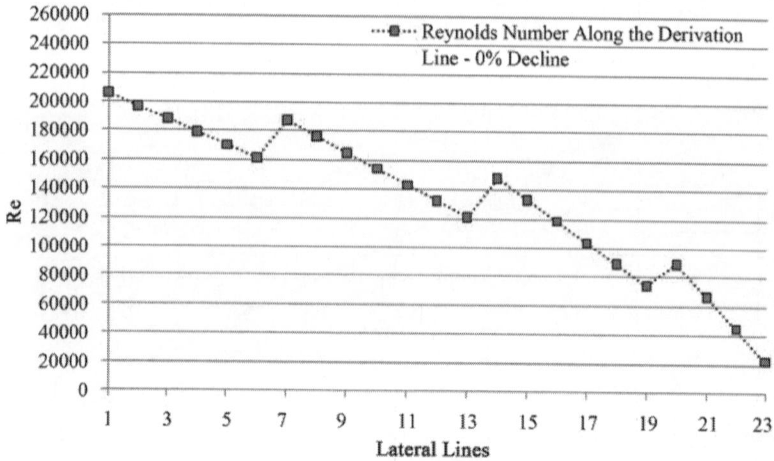

Figure 9. The variation in the Reynolds along the derivation line under a distributed flow with a slope of 0%.

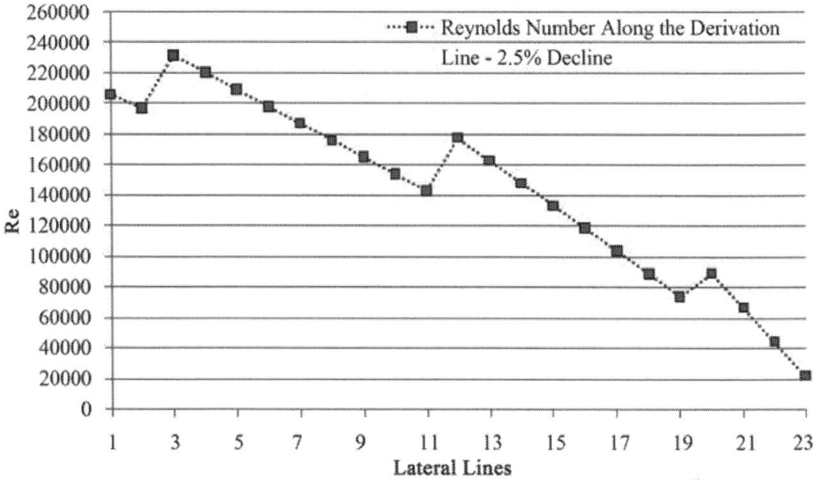

Figure 10. The variation in the Reynolds along the derivation line under a distributed flow with a slope of 2.5%.

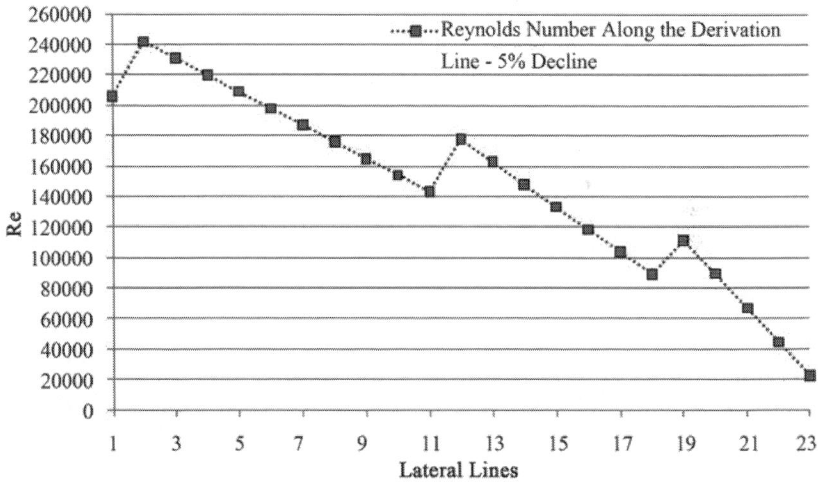

Figure 11. The variation in the Reynolds along the derivation line under a distributed flow with a slope of 5%.

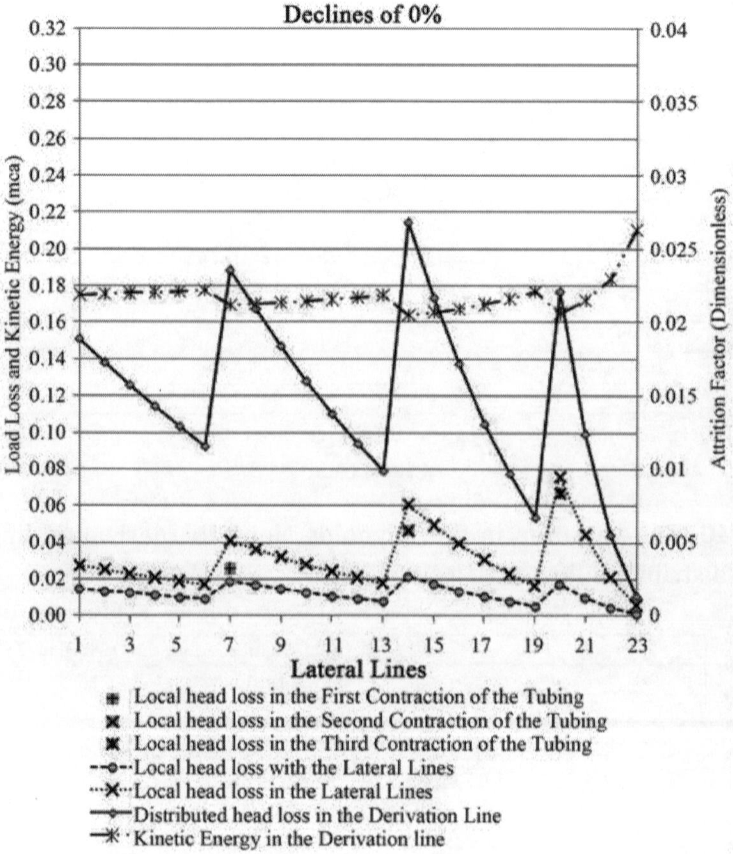

Figure 12. The local head loss in the first, second and third contractions of the tubing and the behaviour of the local head loss in the lateral lines and the head loss unit between the lateral lines and the friction factor and kinetic energy in the derivation line under a distributed flow. Data were obtained from the dimensions of the network with a 0% slope.

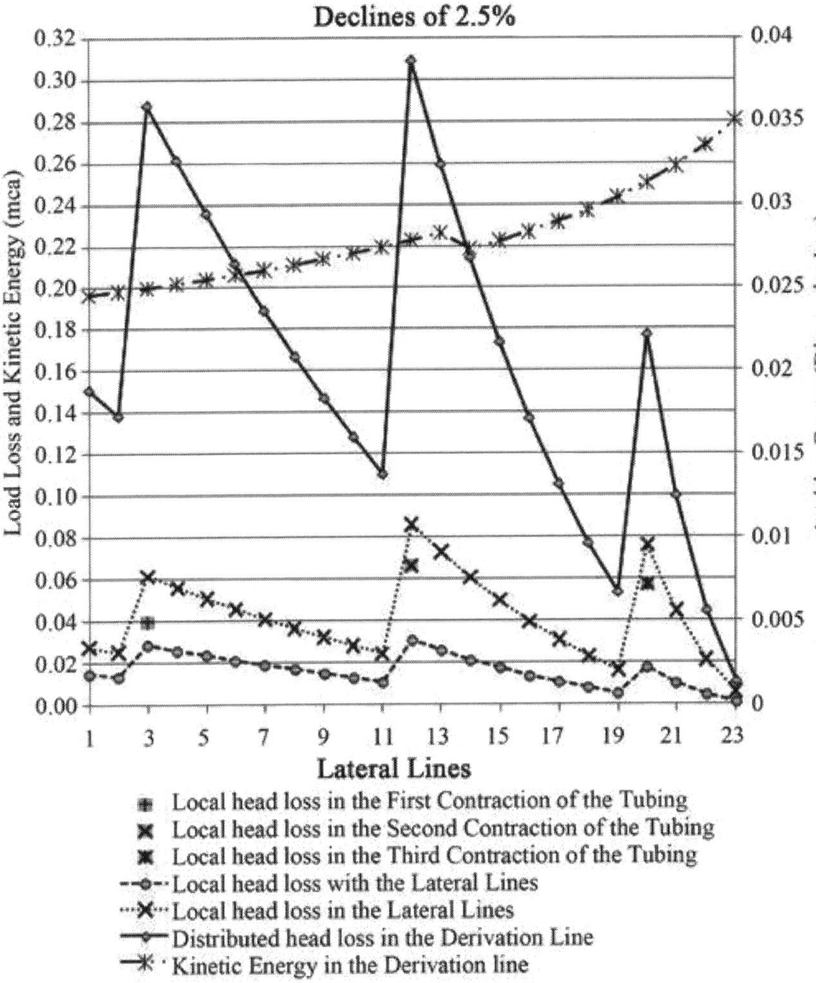

Figure 13. The local head loss in the first, second and third contractions of the tubing and the behaviour of the local head loss in the lateral lines and the head loss unit between the lateral lines and the friction factor and kinetic energy in the derivation line under a distributed flow. Data were obtained from the dimensions of the network with a 2.5% slope.

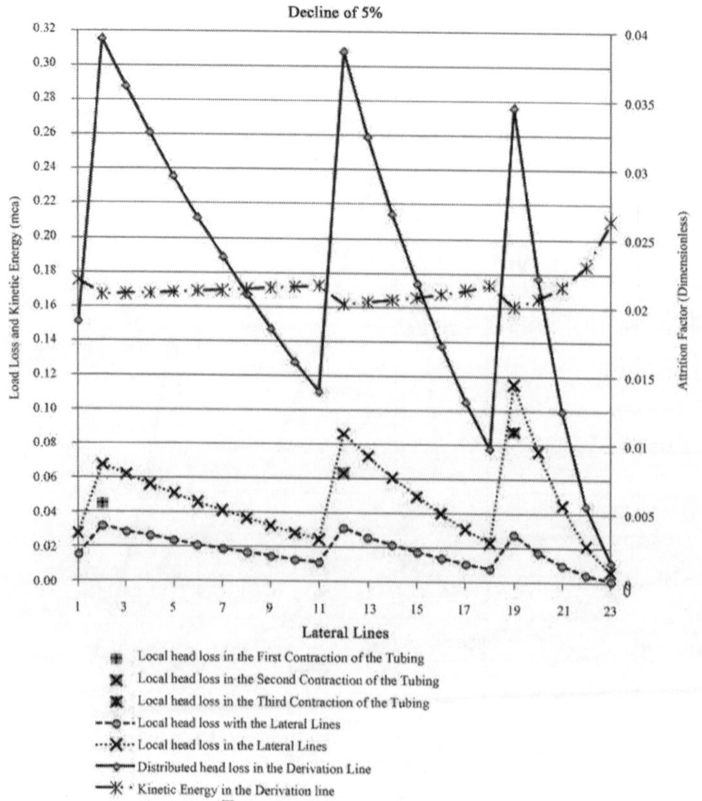

Figure 14. The local head loss in the first, second and third contractions of the tubing and the behaviour of the local head loss in the lateral lines and the head loss unit between the lateral lines and the friction factor and kinetic energy in the derivation line under a distributed flow. Data were obtained from the dimensions of the network with a 5% slope.

slopes, respectively), with the exception of the third contraction of the tubing for the 2.5% slope, which displayed a lower value that that of the 5% slope.

As shown in Figures 12-14, there was less variation in all of the parameters studied for a slope of 0%, with a considerable increase in the variation of the values for slopes of 2.5% and 5%. This shows that as the slope of the terrain increased, all of the parameter values tended to increase, including the amplitude of the local head loss with a contraction in the tubing and the attachment of the lateral lines and the head loss unit between the lateral lines and the friction factor and kinetic energy in the derivation line under a distributed flow.

Figures 15-17 present the energy line, the piezometric line, the pressure head line and the elevation of the derivation line under a distributed flow.

For a slope of 0%, 2.5% or 5%, the kinetic energy (Figures 15-17) in the tubing contributed to the distance between the energy line and the piezometric lines at the beginning of the 4 stretches of the derivation line. At the ends of the four stretches of the tubing, the values of the energy line and of the piezometric line were very close because of the contraction of the tubing, and they were essentially equal at the end of the fourth stretch, when the value of the kinetic energy was close to zero. Additionally, in the comparative analysis of Figures 9-11, the 2.5% slope favoured a better distribution of the energy line and the piezometric line in the tubing because of the gain in energy, which was offset by the total head loss in the tubing.

Figure 16 and Figure 17 show the influence of the elevation and the respective gain in energy (with the tub ing at a downward slope)

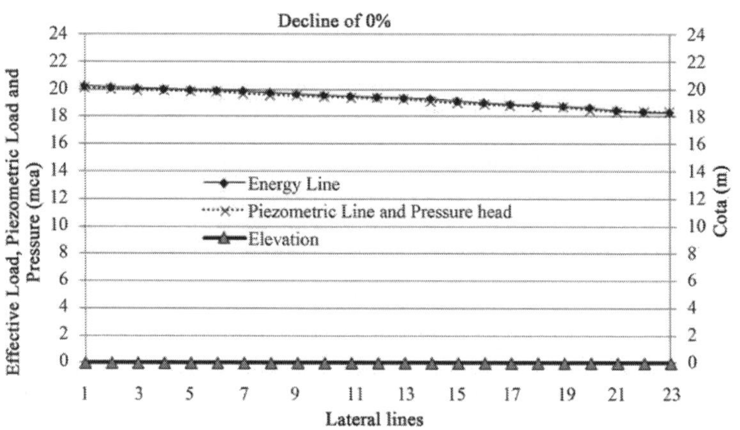

Figure 15. The variation in the energy line, the piezometric line, the pressure head and the elevation of the derivation line under a distributed flow for a 0% slope.

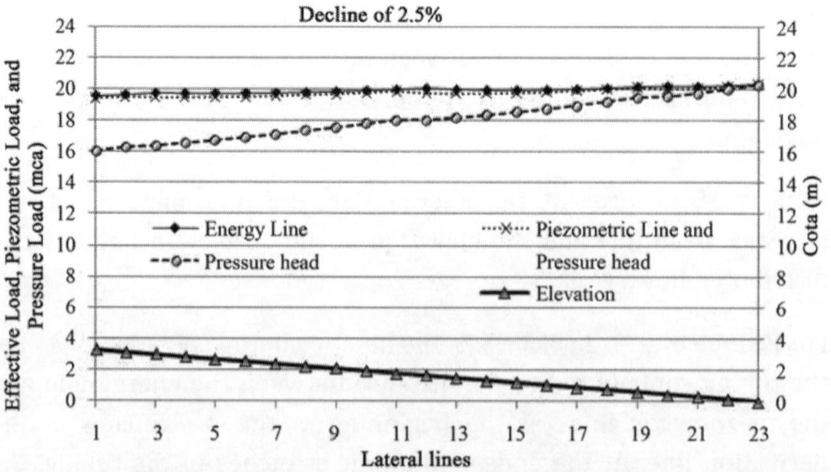

Figure 16. The variation in the energy line, the piezometric line, the pressure head and the elevation of the derivation line under a distributed flow for a 2.5% slope.

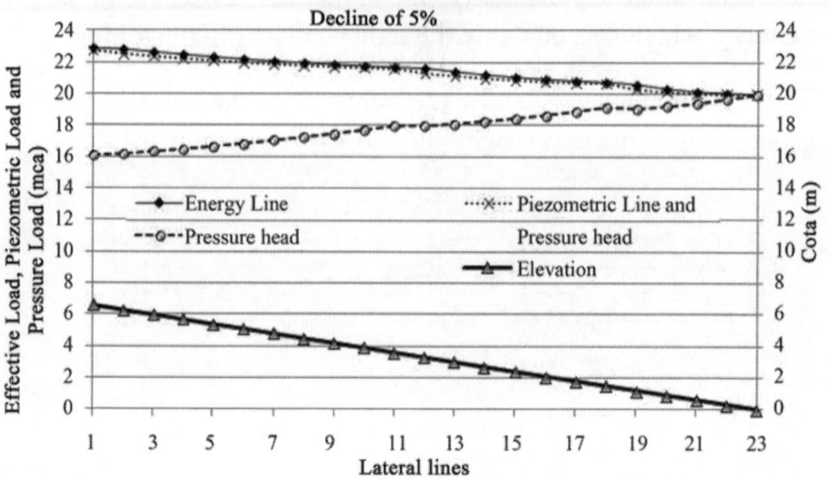

Figure 17. The variation in the energy line, the piezometric line, the pressure head and the elevation of the derivation line under a distributed flow for a 5% slope.

on the energy, piezometric and pressure headlines. This influence determined the distributions in the entire operational unit and, consequently, the flow distribution and the production distribution of the entire area. The clearly and detailed indicate the relationship of each part can be find in [20] .

4. CONCLUSIONS

The results of this study suggested that there was a gain in total head at the beginning of the tubing with an increase in the decline of the terrain (for tubing with a distributed flow) and that the gain was gradually lost along the tubing. There was also a decrease in the pressure at the beginning of the tubing, which gradually increased until the end of the tubing.

From these results, it can be concluded that the lowest average velocity (1.49 $m \cdot s^{-1}$) was obtained in the tubing with a 0% slope, followed by the tubing with 2.5% (1.71 $m \cdot s^{-1}$) and 5% (1.80 $m \cdot s^{-1}$) downward slopes. Thus, there tended to be an increase in the average velocity of the tubing of the derivation line when there was an increase in the decline under the same distributed flow.

The results also showed that for tubing with a distributed flow, the influence of the behaviour of the kinetic energy in the tubing increased the distance between the energy line and the piezometric line at the beginning of each stretch after a decrease in the diameter of the tubing. At the end of the tubing, the values of the energy line and the piezometric line were very similar, and they were essentially equal at the final opening.

The hydraulic analysis showed energy and pressure behaviours that are appropriate for water distribution networks with multiple derivations. The analysis criteria, in relation to the differentiation of the components of the energy line and the piezometric line, facilitated a better hydraulic understanding of the network and the verification of the behaviour of the system, while simulating its operation.

The genetic algorithm used in this hydraulic optimisation resulted in a satisfactory resolution of the problem, without making it necessary to introduce simplifications of the dimensions in the formulation of the code for the capability function.

ACKNOWLEDGEMENTS

Francisco Fernando Noronha Marcuzzo, Edson Cezar Wendland (processes 141159/2006-5 and 504892/ 2008-8) for the financial support that made this study possible.

REFERENCES

1. Kuwabara, G. and Matsura, E.E. (1996) An Evaluation of the Hydraulic Behaviour of the Lateral Lines of Dripper Tubes. In: National Congress OF Irrigation and Drainage, 11. 1996, Campinas. Anais... Viçosa: Brazilian Association of Irrigation and Drainage, 275-287.

2. Santiago, F.S., Montenegro, A.A.A. and Montenegro, S.M.G.L. (2004) Evaluation of Hydraulic Parameters and the Management of Microsprinkler Irrigation in a Settlement Area. Engenharia Agrícola, 24, 632-643.

3. Souza, I.H., Andrade, E.M. and Silva, E.L. (2005) Design and Hydraulics Evaluation of a Low-Head Microirrigation Bubbler System. Engenharia Agrícola, 25, 264-271.

4. Saad, J.C.C. and Mariño, M.A. (2002) Optimum Design of Microirrigation Systems in Sloping Lands. Journal of Irrigation and Drainage Engineering, 128, 116.

5. Saad, J.C.C. and Frizzone, J.A. (1996) Design and Management Optimization of Trickle Irrigation System Using Non-Linear Programming. Journal of Agricultural Engineering Research, 64, 109-118.

6. Lucena, K.F.M. and Matos, J.A. (2001) Economic Analysis of Sub-Units of Localised Irrigation. In: Brazilian Congress of Agricultural Engineering, 30, 2001. Foz do Iguaçu: Anais... Foz do Iguaçu: Brazilian Association of Agricultural Engineering, 2001. 1 CD-ROM.

7. Sarkar, S., Goswami, S.B., Mallick, S. and Nanda, M.K. (2008) Different Indices to Characterize Water Use Pattern of Micro-Sprinkler Irrigated Onion (Allium cepa L.). Agricultural Water Management, 95, 625-632.

8. Sezen, S.M., et al. (2011) Comparison of Drip and Sprinkler Irrigation Strategies on Sunflower Seed and Oil Yield and Quality under Mediterranean Climatic Conditions. Agricultural Water Management, 98, 1153-1161.

9. Gomes, A.W.A., Frizzone, J.A., Rettore Neto, O. and Miranda de, J.H. (2010) Localised Head Loss in Drippers Integrated into Polyethylene Tubes. Scielo Web Publishing.http://www.scielo.br/scielo.php?script=sci_arttext&pid=S0100-69162010000300008&lng=pt&nrm=iso

10. Zitterell Danieli, B., et al. (2009) Head Loss of Microtubes and Connectors Used in Irrigation. Publishing Scielo Brasil. http://www.scielo.br/scielo.php?script=sci_arttext&pid=S0100-69162010000300008&lng=pt&nrm=iso

11. Jin, X., Zhang, J., Gao, J. and Wu, W. (2007) Multi-Objective Optimization of Water Supply Network Rehabilitation with Non-Dominated Sorting Genetic Algorithm-II. Journal of Zhejiang University SCIENCE A, 9, 391-400.http://www.springerlink.com/content/e21522736v131826/fulltext.pdf

12. Opan, M. (2010) Irrigation-Energy Management Using a DPSA-Based Optimization Model in the Ceyhan Basin of Turkey. Journal of Hydrology, 385, 353-360.

13. Singh, R.M. (2010) Design of Barrages with Genetic Algorithm Based Embedded Simulation Optimization Approach. Water Resources Management, 25, 409-429.http://www.springerlink.com/content/d276u4327rr80n80/fulltext.pdf

14. Scaloppi, E.J. (1988) Adjusted F Factor for Multiple-Outlet Pipes. Journal of Irrigation and Drainage Engineering, 114, 169-174.

15. Swamee, P.K. (1993) Design of a Submarine Pipeline. Journal of Transportation Engineering, 119, 159-170.

16. Porto, R.M. (2006) Basic Hydraulics. 4th Edition, Rettec Gráfica and Editor, São Paulo.

17. Aneel, National Agency of Electric Energy (2006) Homologation Resolution No. 313 on the Fees for Providing Elec- tric Energy. ANNEL Web Publishing, Brasilia, 12 p.http://www.aneel.gov.br/cedoc/reh2006313.pdf

18. Noronha, J.F. and Latapia, M.X.I.C. (1998) Custos de produção sob condições de risco no estado de São Paulo. Revista de Economia e Sociologia Rural, 26, 275-287.

19. Frizzone, J.A. (2005) Análise de decisão econômica em irrigação. ESALQ, Piracicaba, 371 p.

20. Marcuzzo, F.F.N. (2008) Sistema de otimização hidráulica e econômica de rede de irrigação localizada usando algoritmos genéticos. Tese de doutorado. EESC/USP, São Carlos, 361 p. http://www.teses.usp.br/teses/disponiveis/18/18138/tde-11112008-091041/pt-br.php

21. Marcuzzo, F.F.N. and Wendland, E. (2010) Efeito da variação na tarifação pelo uso da água no dimensionamento otimizado de rede de irrigação localizada usando algoritmos genéticos. Revista Brasileira de Recursos Hídricos, 15, 109-118.https://www.abrh.org.br/sgcv3/UserFiles/Sumarios/09

65dabc71f46b05537f68871d5ed280_de994387275c4b2e63 27350b0de23f65.pdf

22. Marcuzzo, F.F.N. and Wendland, E. (2010) Otimização de rede de irrigação de microaspersão usando algoritmos genéticos sob diferentes declividades e tarifação de água e energia elétrica. Engenharia na Agricultura, 18, 50-62.http://www.seer.ufv.br/seer/index.php/reveng/issue/view/12/showToc

23. Marcuzzo, F.F.N. and Wendland, E. (2011) Distribuição de pressão em rede de irrigação localizada otimizada por algoritmos genéticos. Engenharia Agrícola, 31, 497-505.

9

PHOTOVOLTAIC WATER PUMPING SYSTEM IN NIGER

Madougou Saïdou[1], Kaka Mohamadou[1] and Sissoko Gregoire[2]

[1] Université Abdou Moumouni de Niamey, Niger

[2] Université Cheikh Anta Diop de Dakar, Senegal

1. INTRODUCTION

Solar energy such as photovoltaic is the most promising energy of the non-conventional energy sources which is capable to satisfy the energy needs of the isolated rural areas. It fits perfectly to the decentralization of power generation for the small communities widely dispersed as evidenced the solar pumps whose operation is nowadays proved flawless (World Resources Institute, 1992).

The use of photovoltaic energy to pump water is particularly well suited in the Sahel. This source of energy is free and abundant, but also provides autonomy for many isolated villages of rural areas. The water pumped is stored in the thanks until its use (in the night or during the cloudy days). Locally, there are many companies that manufacture these tanks. These towers do not require special maintenance and are easy to be repair. This chapter presents a study of photovoltaic water pumping in Niger rural areas. We present first the benefits of photovoltaic water pumping; we describe in the second the photovoltaic water pumping system, before present the solar radiation at Niger. Then the sizing of a photovoltaic water pumping system is described and finally, a case study of photovoltaic water pumping is presented before concluding.

2. BENEFITS OF PHOTOVOLTAIC WATER PUMPING SYSTEM

In African countries, many governments are still struggling to meet the basic needs of the population due to lack of availability of electric power system. In rural areas, these needs are summarized in the drinking water, electrification of health centers and irrigation. For lack of resources, these people are resolved to supply water from wells craft of 10 to 100 m and less than 2 m of diameter. Water is the source of life. Its rarity is one of the tragedies of the Sahel. In this area, people have regularly serious problems of drinking water. The water which involves in agricultural and domestic consumption requires dewatering technologies adapted to local conditions. Yet, the solar energy potential is very abundant. The geographical situation of Niger fosters the development and growth of the use of solar energy. Indeed, given the importance of the solar radiation and the duration of sunshine, which excesses eight hours (8 hours) per day throughout the year, some electricity needs can be met by solar energy. This energy source may be beneficial to the most remote areas, especially in water pumping. The use of solar energy can make a meaningful and lasting solution to the drainage in rural areas. The use of photovoltaic solar energy for water pumping is well suited to these regions due to the existence of a potential groundwater quite important especially in desert areas, and a large solar energy potential. Also, this solar energy has the advantage of being clean compared to conventional energy sources which present constraints of remoteness from the mains, fuel transport and periodic maintenance for diesel engines (Jimmy et al 1998, Barlow, 1993). Photovoltaic solar energy may well contribute to meet the energy needs of these populations. Photovoltaic generators are very suitable to relieve the lack of availability of grid. Indeed, with an installation of 1 kW maximum power, we can pump water to heights of about 80 m, or meet irrigation needs of a village (Commission of The European Community, 1985). In addition, these solar generators can be used concomitantly using batteries to power television sets, or for lighting when the pumping system is not loaded (Billerey, 1984).

3. PHOTOVOLTAIC PUMPING SYSTEM

3.1. Description

Photovoltaic pumping system is a standard pump equipped with an electric motor, provided in electrical energy by photovoltaic panels installed on the site (Handbook on Solar Water Pumping, 1984; Fraenkel and al., 1986). This pump is intended to pump water from the basement to make it accessible to users (Fig. 1).

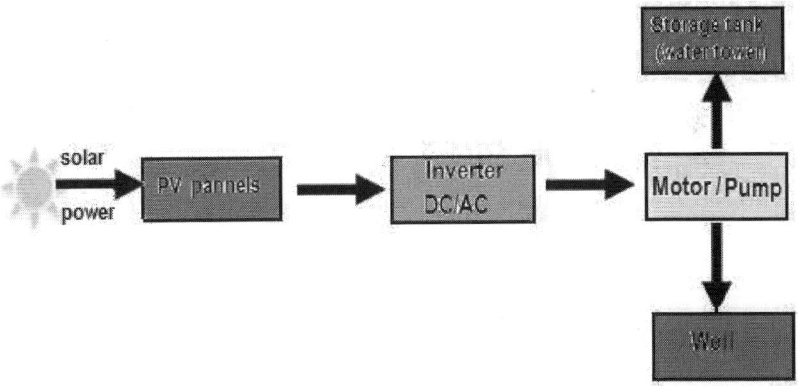

Figure 1. Principe of photovoltaic pumping system.

Nowadays, two types of photovoltaic water pumping systems are used: the photovoltaic water pumping with batteries and without batteries. In Niger, it is often used photovoltaic water pumping without batteries, commonly known as "pumping over the sun". Pumping over the sun is simpler and less expensive than with battery system. Instead of batteries, they use a tank to store water until it is used (Fig.2). Hydraulic storage allows overcoming electrical energy storage thus avoiding the use of batteries which have a limited life (6 years compared with 20 to 30 years of photovoltaic panels) and are polluting. However, the method without batteries has some drawbacks and its main fault is to have a flow of water which depends on the average time of the sunlight (A. Hadj, 1999).

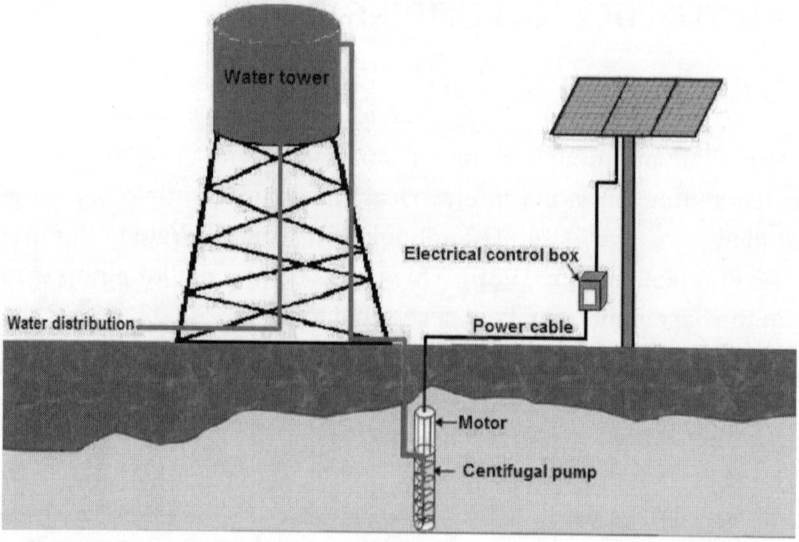

Figure 2. Photovoltaic water pumping without batteries with a tank to store water.

There are two types of pumps to draw water surface: Positive Displacement *Pumps* (volumetric pumps) and centrifugal pumps. Besides to the type of pump, there are two other characteristics at the pumps according to the physical location of the pump in relation to the pumped water: the suction system and stuffer one. They discharge pumps are submerged in water. Their motor is immersed in water with the pump and the discharge pipe placed after the pump can lift water to tens of meters to the storage tank depending to the engine's power. Afterward, the system is connected to a distribution network that delivers water to users.

3.2. Photovoltaic System

To generate the necessary energy to the motor of the pump, solar photovoltaic panels are placed for converting solar energy into electrical energy (Fig.3). As the panels generate a direct current (DC), it is often used DC/AC converter to convert the direct current produced by the solar panels into alternative current (AC) if the motor of the pump is AC. On the other side, if the motor is DC, the device does not need a DC/AC converter. The energy produced by the panels can be used directly or stored. In the case of an application for water pumping, it is more interesting to use the energy to raise the water in a castle that serves as hydraulic energy storage.

Figure 3. Students determining the current-voltage characteristic I = f(V) of solar cell "over the sun" at the University Abdou Moumouni of Niamey (Niger).

To prevent a dysfunction of the pump when it is live on photovoltaic, due to under sizing or over sizing the PV generator, an inverter is used to ensure the proper operation of the PV/pump system.

4. SOLAR RADIATION AT NIGER

To design a photovoltaic water pumping system, we will need to quantify the available solar energy. Therefore, it is very important to know the solar radiation of the locality. Solar radiation (kW/m^2) is the energy from the sun that reaches the earth. The earth receives a nearly constant of solar radiation at its outer atmosphere. The intensity of solar radiation varies with geographic location (Fig.4).

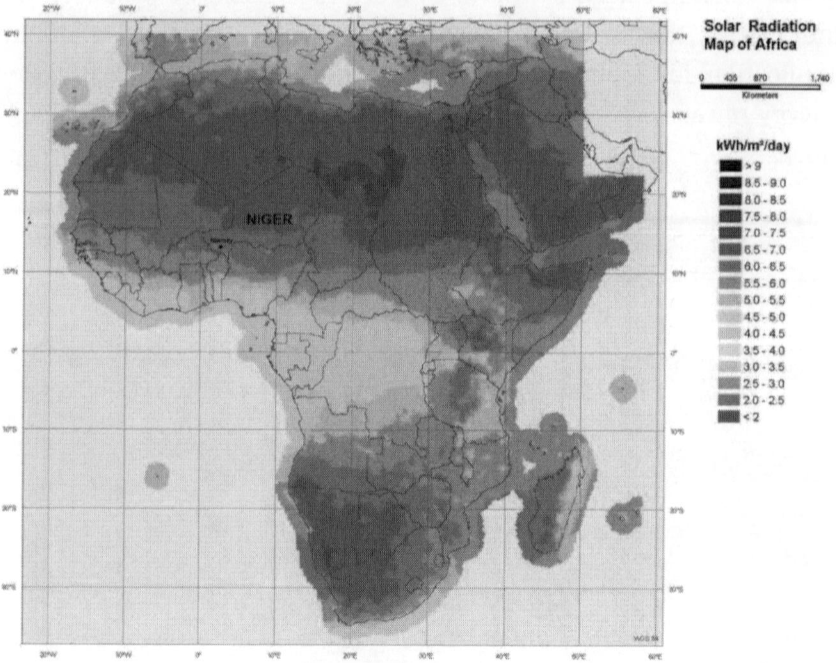

Figure 4. Africa solar radiation map (Source: UNEP, NREL and the Global Environment Facility).

It also varies with the season and time of the day. In Sahel, the solar radiation at the earth's surface is very important (Fig.5).

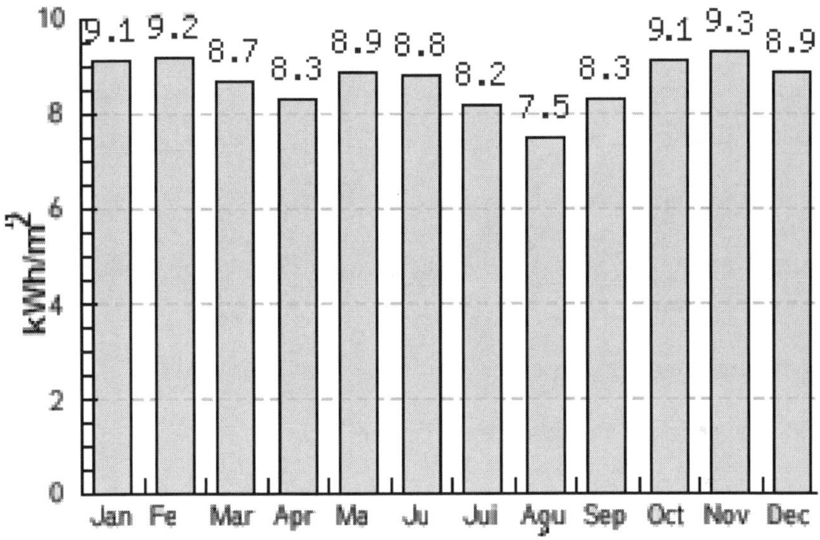

Figure 5. Daily average radiation of sunlight per month at Niger (kWh/m^2.day).

The most productive hours of sunlight are from 9:00 a.m. to 5:00 p.m. Table 1 gives the daily average time (Hours/day) of the sunlight at Niger (Keita town: Latitude 14.75°N and Longitude 5.76°E).

Table 1. Monthly average time (Hours/days) of the sunlight at Niger (Keita town: Latitude: 14.75°N and Longitude: 5.76°E)

Month	Jan.	Feb.	Mar.	Ap.	Ma	June	July	**Aug.**	Sept.	Oct.	Nov.	Dec.
Average Time (Hours/day)	9.0	9.3	8.7	8.4	8.8	8.6	8.2	**7.5**	8.1	9.2	9.5	8.9

5. Sizing a photovoltaic water pumping system

Photovoltaic water pump sizing is the determination of the power of the solar generator that will provide the desired amount of water (Alonzo, 2003).

The photovoltaic water pump sizing consists of:

- Assessment of daily water needs of the population to know the rate flow required;
- calculation of hydropower helpful;
- determining of the available solar energy;
- determining of the inclination of the photovoltaic generator which can be placed;
- determination of the month sizing (the month in which the ratio between solar radiation and hydropower is minimum);
- sizing of the PV generator (determination of the required electrical energy);

5.1. Assessment Of Daily Water Needs Of The Population

Determining the water needs of the consumption of a population depends mainly of its lifestyle, the environment and climatic conditions of each region. Drinking, cooking, washing and bathing are the main uses of water for human needs. Animals also need water for their survival. The water use is also essential in the field of agriculture.

Depending on the nature of the users (humans, animals) or the use, the amount of water required for each user or usage are:

- 30 liters /day, the normal living conditions in Africa;

- **Animals**

- Cattle	40 liters /day;
- Sheep, Goat	4 liters /day;
- Horse	40 liters /day
- Ass	20 liters /day
- Camel	20 liters /day (reserve for 8 days)

- **Irrigation (agriculture)**

- Crops at the village	60 m^3 /day / hectare;
- Rice	100 m^3 /day / hectare;
- Cereals	45 m^3 /day / hectare;
- Sugar cane	65 m^3 /day / hectare;
- Cotton	55 m^3 /day / hectare.

There are three standards for the calculation of water requirements:

1. The standard for the minimum amount for survival.
2. The current target of funding agencies: 20 liters/day/person that does not include livestock and gardening.
3. The minimum amount necessary for economic development: 50 liters/day/person. It includes:

- 20 liters/day/person, for personal needs;
- 20 liters / person / day, 0.5 unit of cattle per person;
- 10 liters / day / person, 2 m^2 for vegetable crops.

In applying the standards 2 and 3, we obtain the following water requirements (Table 2) depending on the size of the village (population).

5.2. Determination Of Hydropower Helpful

The average daily load i.e. hydropower helpful (kWh/day) required is expressed by:

$$E_H = \frac{g^*\rho_a^*Q_a^*TH}{\eta_P^*3600} = \frac{C_H^*Q_a^*TH}{\eta_P} \quad (1)$$

Table 2. Water requirements depending on the size of the village (population).

Population of village (persons)	Standard 2: (20 liters/day/person) (m^3)	Standard 3: (50 liters/day/person) (m^3)
250	5	12.5
500	**10**	**25**
750	15	37.5
1000	20	50
1500	20	75
2000	40	100

Where, ggis acceleration of gravity (9.81 m.s^{-2});

ρapρapais water density (1000 kg/m^3);

Qa Qais daily water needs (m^3/day);

TH TH is the total head (m);

ηPηPis pump system efficiency

The tank capacity is determined by the daily water needs and the autonomy of the system.

5.3. Determining Of The Available Solar Energy

The method used is based on the determination of daily mean values of solar radiation available and hydropower necessary.

5.4. Determining Of The Inclination Of The Photovoltaic Generator Which Can Be Placed

The inclination ββ to the horizontal plane of the photovoltaic panels (PV) must be to maximize the relationship between solar radiation and hydropower necessary. We have chosen 15°N, the latitude of the locality (Keita town: Latitude 14.75°N).

5.5. Determination Of The Month Sizing

The sizing month will be the worst month, i.e., the month that the ratio between solar radiance and hydraulic energy required is minimal. In our case it is the month of August is the month of sizing. In august the average time is 6.5 hours/day.

5.6. Sizing Of The Pv Generator

As the system works all year round, the field is tilted at an angle equal to the latitude of 15 ° N. It was in August that the average number of hours of sunshine is the lowest maximum: 6.5 hours/day. Assuming a 25% loss due to the temperature and dust, the required electrical energy is given by:

$$W_{PV} = \frac{E_H}{Radiance*(1 - loss)} \tag{2}$$

6. CASE STUDY

A study on the photovoltaic water pumping system in a village at 30 km of Keita (Niger) to meet the water needs of the five hundred (500) persons gave the following summarized results.

Figure 6. Photovoltaic water pumping station with a volumetric pump at Niger.

a. Assessment of daily the water needs: Using the Standard 3: (50 liters/day/person), the water needs rises to 25 m³/day.

b. The average daily load i.e. hydropower helpful (kWh/day) required is given by this expression :

$$E_H = \frac{g^* \rho_a{}^* Q_a{}^* TH}{\eta_P{}^* 3\,600} = \frac{C_H \cdot Q_a{}^* TH}{\eta_P}$$

With $g = 9.81\ \mathrm{m.s^{-2}}$

$\rho_a = 1000\ \mathrm{kg/m^3}$

$Q_a = 25\ \mathrm{m^3/day}$

$TH = 52\ \mathrm{m}$

$\eta_P = 50\ \%$

It provides: $E_H = 7\,085\ \mathrm{Wh}$

a. The available solar energy:

b. Daily average radiation of sunlight varies from 7.5 to 9.3 kWh/m²/day.

c. To make sure to do a good sizing, we choose the minimum value of average radiance: 7.5 kWh/m2/day.

The inclination to the horizontal plane of the photovoltaic panels is: β=15°Nβ=15°N.

a. The sizing month is: August, 6.5 hours/day.

b. Sizing of the PV generator

Assuming a 25% loss due to the temperature and dust, the required electrical energy is given by this expression:

$$W_{PV} = \frac{E_H}{Radiance^{*}(1 - loss)}$$

$$W_{PV} = 1260 \text{ Wc}$$

The operating point of our photovoltaic field is set at 120 volts due to the characteristics of the inverter. The photovoltaic field will be composed of 10 multiple modules in series. Generator power is 1260 Wc, the rate current is 10.50 A. With photovoltaic panels which have 3.5 A, we will have 3 modules in parallel.

The table 3 shows the summary of case study.

Table 3. Summary of case study

Month of sizing	Radiance (kWh/m²)	Time of sunlight (H)	Loss (%)	E_H(kWh)	W_{PV}(Wc)	Voltage (V)	Current (A)	Configuration of panels
August	7.5	7.5	20	7 085	1260	120	3.5	10 X 3

7. CONCLUSION

The use of solar energy in Niger, particularly photovoltaic energy, for water pumping is well suited in this arid and semi-arid area due to the existence in this region of an underground water potential, and a large solar energy potential more than 6 kWh/m^2.

Photovoltaic generators are coupled directly to the pump with a DC/AC converter. Storing water in the tanks avoids additional costs accumulator used to store electrical energy. The case study clearly shows the advantage of photovoltaic pumping system compared to conventional energy one which has many constraints of distance to the power grid, of transportation of fuel, and of periodic maintenance of the engines. The cost of one cubic meter (1 m^3) of water pumped by the PV system is more advantageous than others systems. This pumping system constitutes a solution for the water supply of these sparsely populated, remote and isolated areas (Thomas, 1987). With the falling prices of solar panels, this source of energy must be popularized and integrated in the development strategy of these countries.

REFERENCES

1. A Hadj, F Arab, K Chenlo, Mukadam and J.L. Balenzategui, (1999Performance of PV Water Pumping Systems". Renewable Energy, 18N°2, 191204

2. R Barlow, B Mcnelis, Et Derrick, A, (1993Solar Pumping: An Introduction and Update on the Technology, Performance, Costs, and Economics, World Bank Technical Paper 168Intermediate Technology Publications, London, UK.

3. J Billerey, 1984Le pompage photovoltaïque, GRET, Paris, France.

4. Commission of The European Community1985Renewable Sources of Energy and Village Water Supply in Developing Countries. Directorate General for Development, EEC.

5. C Alonzo, 2003Contribution à l'Optimisation, la Gestion et le Traitement de l'Energie ». Thèse de Doctorat, Toulouse.

6. P Fraenkel, 1986Water Pumping Devices, I.T Power, London, UK.

7. Handbook on Solar Water Pumping (1984Intermediate Technology Power & Partners, eading, 124 p.

8. R Jimmy, D Thomas, E Schiller, and B. S Sy, 1998Le pompage photovoltaïque : manuel de cours à l'intention des ingénieurs et des techniciens. Éditions Multi Mondes. 254 p. 2-89481-006-7

9. M. G Thomas, 1987Water Pumping- The Solar Alternative". Photovoltaic Design, Assistance Center. Sandia National Laboratories. NM 87185, Albuquerque, 58 p.

10. UNEPNREL and the Global Environment Facility (2012www.unep.org/dgef

11. World Resources Institute1992Ressources mondiales 1992-1993 : un guide de l'environnement global, PNUD, Éditions Sciences et Culture, Montréal, Canada.

10

DEVELOPMENT OF A PORTABLE ELECTRO-MECHANICAL EDUCATIONAL MODEL FOR VARIABLE RATE CENTER PIVOT IRRIGATION TECHNOLOGY

Young J. Han[1], Ahmad Khalilian[2*], Jose Payero[2], Nicholas Rogers[1]

[1]Department of Agricultural and Environmental Sciences, Clemson University, Clemson, SC, USA

[2]Edisto Research and Education Center, Department of Agricultural and Environmental Sciences, Clemson University, Blackville, SC, USA

ABSTRACT

Center pivot irrigation systems usually apply a relatively uniform amount of water to fields that are often inherently variable, which could lead to significant waste of water and energy. To address this issue, our team is now developing an Intelligent Center Pivot (ICP) by integrating sensor-based irrigation scheduling with variable rate irrigation technology. However, before this technology can be applied in commercial production, it is necessary to educate growers about its practicality and potential benefits. The objective of this study was to develop a portable tabletop intelligent center pivot model (ICPDemo) to demonstrate and promote adoption of the ICP technology. This paper describes an ICPDemo constructed in 2014, including the design specifications, electro-mechanical design, control strategy, and performance. The ICPDemo has performed according to design specifications and is successfully being used to demonstrate the benefits and effectiveness of ICP technology for irrigation scheduling.

Keywords: Irrigation, Center Pivot Irrigation, Electro-Mechanical Model, Sensors, Control, Variable Rate Irrigation

1. INTRODUCTION

Center pivot irrigation systems are predominantly designed to apply a relatively uniform amount of water to fields that are often inherently variable due to factors such as different soil types, topography, or multiple crops. This field variability can result in significant waste of water and energy, runoff, chemical leaching, and reductions in yields and farm profits. Variable-rate irrigation has the potential to improve the efficiency of water and energy use as well as crop yields. Variable rate center pivots have been studied by [1] and [2] for their advantages compared with single rate irrigation systems. Over the last decade, the Precision Farming Team at Clemson University has developed a variable-rate lateral move irrigation system that has been successfully deployed by growers in commercial production in South Carolina [3] - [5] . However, since center pivot irrigation is the predominant irrigation method for row crop production in South Carolina, there is a need to refine the technologies developed for the lateral irrigation system to include center pivot irrigation systems [6] and [7] . Therefore, current efforts are being directed towards developing an Intelligent Center

Pivot (ICP), which will integrate sensor-based irrigation scheduling with variable rate irrigation technology.

In order to reduce waste of water and nutrients and improve crop yields through this technology, there has to be an understanding among potential users about the technical aspects and potential cost/benefits of this technology. In the past, electro-mechanical models have been used successfully to demonstrate irrigation technology. For example, in 2010, our research team at Clemson University developed a portable electro-mechanical model to demonstrate nozzle pulsing technology for variable-rate lateral irrigation systems. Also, about the same time, researchers at the University of Georgia [8] developed a table-top variable-rate center pivot model to demonstrate nozzle pulsing technology. Both of these systems were successfully used in a variety of irrigation extension activities for several years, but they only focused on demonstrating VRI technology specially on nozzle pulsing technology. Now, since the technology has advanced, there is a need to demonstrated integration of sensor-based irrigation scheduling with VRI technology. Currently, there is no published information related to development and use of this type of models for irrigation demonstration.

A portable electro-mechanical model to effectively demonstrate the concepts of the intelligent center pivot technology to farmers, especially in indoor settings such as trade shows and farmer's meetings, is currently not available. Therefore, the objective of this study was to develop a portable tabletop intelligent center pivot demonstration model (ICPDemo) that could be used to demonstrate and promote adoption of the ICP technology among growers. This paper describes the design and construction of the ICPDemo, including the design specifications, electro-mechanical design, control strategy, and performance.

2. METHODOLOGY

2.1. Design Specifications

The final product should comply with the following specifications:

- Be a fully functional tabletop model of an ICP, which integrates variable rate irrigation system and irrigation scheduling.
- Be less than 82 cm wide in order to be portable and easily fit through a standard door. This would affect: the size of the pivot boom itself (considering length and therefore relative height), the type of irrigation nozzle to allow for the proper throw of water, the size of the pumps to allow for adequate water flow, and the size and design of a reservoir (tank) to hold the water.
- Have at least four management zones plus an end gun whose application rates can be individually controlled over a number of control sections in the field.
- Have a working pump and irrigation nozzles across different management zones whose application rates can be individually controlled according to an irrigation rate map.
- Have an electronic controller with enough controllable input/output channels and enough electrical current capacity to run pumps and motors.
- Have a functional irrigation system with site-specific variable-rate control technology similar to those used in modern precision agriculture.
- Have a rotational pivot speed that can be automatically controlled according to the irrigation rate map. The total run time should be approximately one minute from one end to the other in order to allow for adequate time to demonstrate each management zone.

- Be a foolproof turn-key system that is simple and robust, so that anyone with little training can operate it without consulting a manual.

2.2. Electro-Mechanical Design and Construction

The final design was a small two-section center pivot covering a semi-circular area (Figure 1). The multi-view engineering drawings (front, side, and top views) of the system, including dimensions, are shown in Figures 2-4. The ICPDemo was constructed using 38 mm aluminum tubing for the boom and supports. The design also included a tabletop map with several management zones and different field features, such as various soil types and a standing pond. During construction, five 6.4 mm flexible tubes were routed through the boom from the pumps to the nozzles. This provided water flow to each section of the boom. The pivot was designed to rotate using two small reversible motors (5-rpm, 12VDC). One motor had enough torque to run the pivot, but the second one was added for redundancy. The motor shafts were attached directly to the pivot wheels (10-cmdiameter), which were taken off a small toy tractor. The size of the driving wheel and the rotation speed of the motor were chosen to achieve the total run time of approximately one minute from one end to the other end of the tabletop (0 - 140 degrees of rotation). A manual motor-override switch was installed to stop the system, in case more time for explanation was needed during any demonstration. A main power switch for the whole model and an individual pump override switch for each section of the pivot were also installed.

A total of five small 12VDC pumps, often found in automobile windshield cleaning systems (ACI Windshield Washer Pump, Advance Auto #172435), were used. Four of the pumps were used to control each of the four management zones (2 nozzles per zone), with the fifth pump controlling the end gun. A separate pump was needed for each management zone, in order to control the irrigation rate based on the water requirement of that particular zone. These pumps could be turned on and off to demonstrate the nozzle pulsing technique. Water was stored in a reservoir constructed from clear LEXANTM sheets joint together with LEXANTM cement. This reservoir also allowed for secure mounting of all the pumps (Figure 5).

Figure 1. Final design of intelligent center pivot demonstration model (ICPDemo).

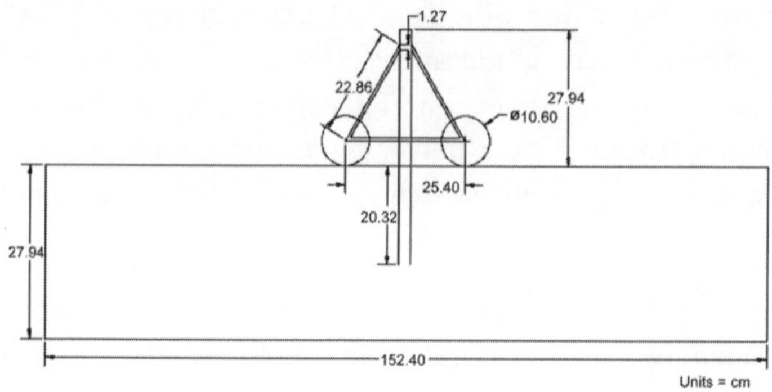

Figure 2. ICPDemo—front view.

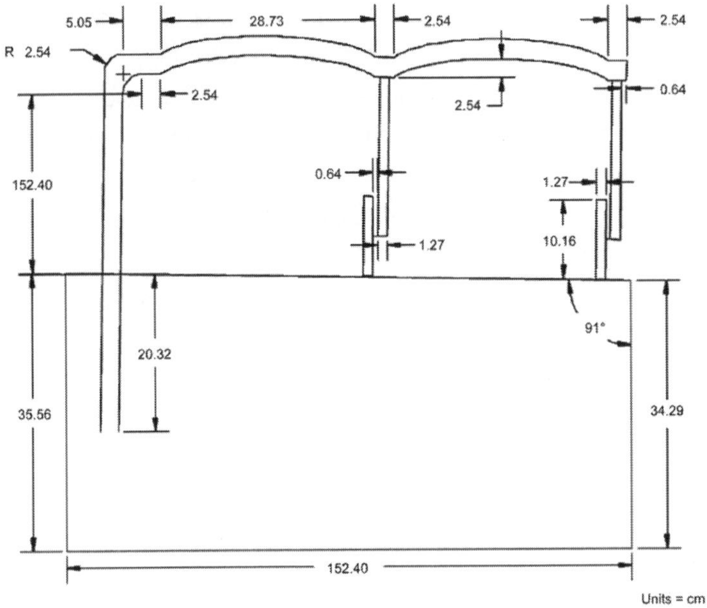

Figure 3. ICPDemo—side view.

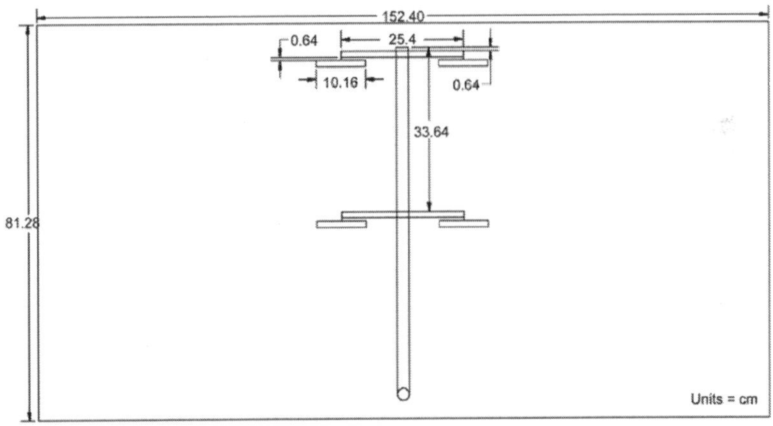

Figure 4. ICPDemo—top view.

The main control box under the ICPDemo was constructed from wooden boards (5 × 10 cm), plywood, and paneling. This box housed the water reservoir, 12V power supply (Pyramid #PSV300), relays, controllers and all wiring. The top of the box was angled on two planes to allow all water applied to drain to one corner. In this corner a hole was cut in the top of the box and a funnel was placed inside to allow all water to drain back into the reservoir to be recirculated. A 25-mm ID flange bearing (Peer Bearings, HCFTS205-16) was mounted to the underside of the top of the box to allow the pivot to rotate with reduced friction. The simulated field map was laminated and placed on top of the box and a piece of clear LEXANTM sheet was also attached on top of the map to cover the full length and width of the box. The top of the box was surrounded by 25-mm walls made of the LEXANTM sheet around the perimeter to ensure all water recirculated properly.

Figure 5. Water reservoir with mounted pumps.

A Programmable Logic Controller (PLC) (Crouzet Automation Model CD-20, Coppell, TX) was used to control the five pumps and the reversible motor. The PLC had 12 inputs and 8 outputs and operated on 12VDC power supply. A PC-based data acquisition and control system was considered, but the PLC was chosen for its simplicity and versatility. Anyone with little training can operate the PLC by simply turning on a switch, without waiting for a PC to boot up and load specialized software.

The full scale VRI pivot normally utilizes a Global Positioning System (GPS) receiver, to locate the position of the pivot within the field for zone management. Therefore, to simulate a GPS, a small Light Emitting Diode (LED, RadioShack, 276-024) was mounted to the end of the irrigation boom and was turned on and off at 1 Hz. Since a GPS receiver would not properly operate indoors, three watertight micro switches (Honeywell Sensing and Control #V15W11-WZ200A06-AW1, Morristown, NJ) were used instead to locate the position of the pivot within the simulated field. The PLC interpreted the output of the micros witches using binary code (open or closed translating to 0 or 1) to provide location information at any given time. For example, the output binary code of "101" indicated that the pivot was in section 5 (Figure 6). A small switch plate with ridges and grooves was constructed to allow the micro switches to open or close at the appropriate times to correspond with each management zone (Figure 6).

Figure 7 shows the wiring schematic of the PLC. The five water pumps (indicated as "P") were wired into the first five outputs channels (01 to 05). The first four pumps controlled one management zone each, and the fifth pump operated the end gun. Two relays connected to the next two outputs controlled the drive system for the pivot. The first relay (T92S11D22-12, Potter & Brumfield) simply turned the drive system motor (M) on or off with any specified duty cycle. This enabled the travel speed of the pivot to be controlled precisely as needed by a pulse-width modulation technique. The second relay (KUHP-11D51-12, Potter & Brumfield) controlled the polarity of the 12 volts DC motor (M) to reverse the travel direction of the ICPDemo when needed. The last output of the PLC controlled the blinking of the LED that simulated a GPS receiver. The ICPDemo was fully tested to assure that all design specifications were met.

2.3. Control Strategy

The ICPDemo had several management zones and field features, such as various soil types, to simulate actual condition of the commercial production fields in southeast USA. These features and management zones programmed on the final model can be seen in Figure 8. However, the program can be changed to simulate other conditions. In the example of Figure 8, the four management zones are numbered Z1 through Z4. The area outside Z4 is to be irrigated by an end gun. The field is divided into eight sections, starting from Section 0 on the left at the 0 degree position, and numbered clockwise to the right up to Section 7. The middle six sections (S1-S6) are irrigated according to an irrigation rate map. Sections 0 and 7 are used as a turning area for the pivot to turn around by reversing the polarity of the driving motors. Management Zones and Soil Features in Figure 8 were designed to encompass several different field attributes in order to demonstrate various functions of a Variable Rate Intelligent Center Pivot. In this example, the clay soil received the irrigation rate that the ICPDemo could deliver at full travel speed without pulsing nozzles (1X in Table 1). For the dry sandy soil, the irrigation rate was doubled (2X) by reducing the travel speed to 50%. For the boggy areas, irrigation rate was reduced to 50% of the clay soil (0.5X) by pulsing nozzles.

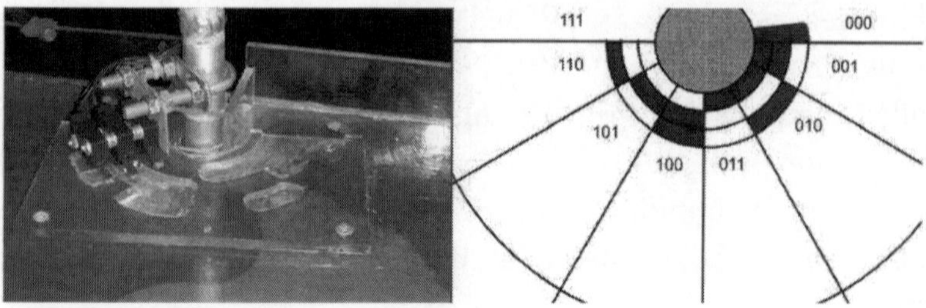

Figure 6. The switch plate with ridges and grooves (left), and output binary codes for each zone (right).

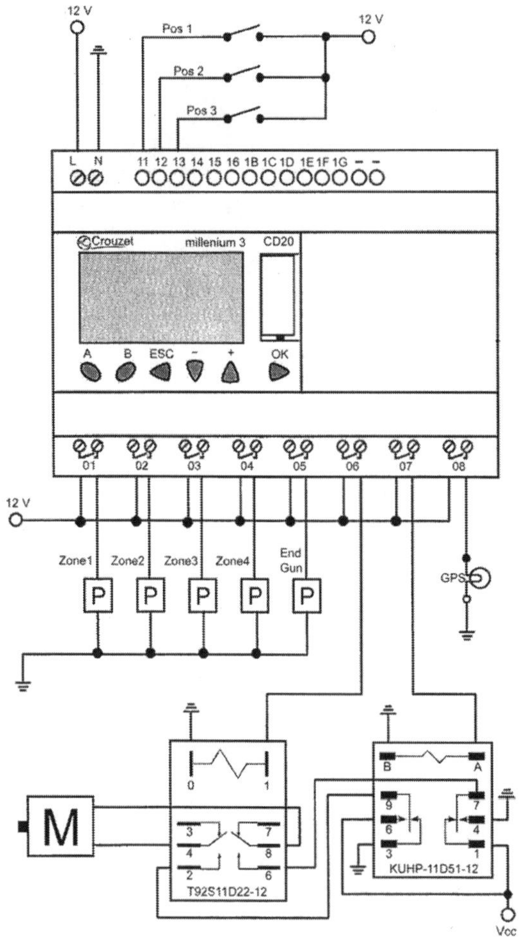

Figure 7. PLC wiring diagram (P = pump; M = motor: T92S11D22-12 and KUHP-11D51-12 = relays; GPS = LED to simulate GPS; and Pos1, 2, 3 = position switches).

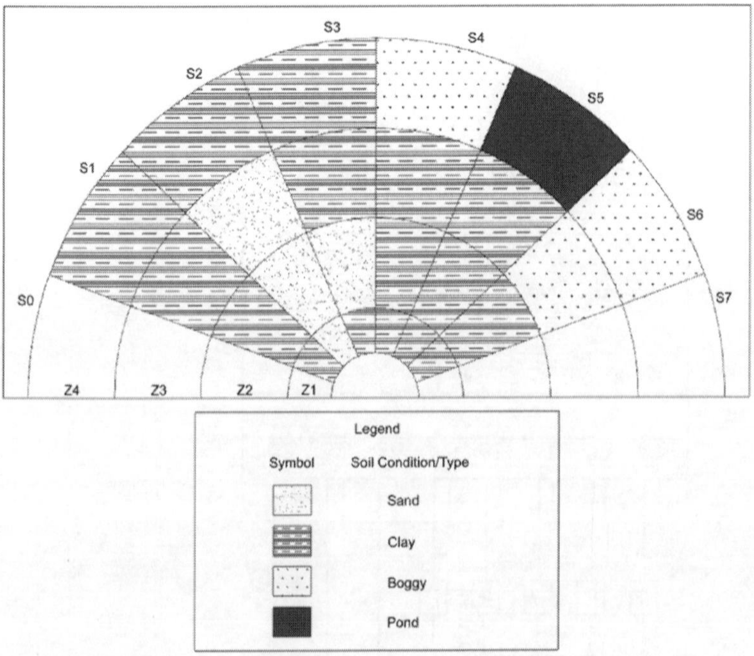

Figure 8. Management zones and soil features.

Table 1 shows the irrigation strategy for each zone through different sections for the given field features. The irrigation rate was controlled by changing the pivot travel speed and/or by pulsing the nozzles. For example, in the first section (S1) where all zones had clay soil type, all four zones would receive the same irrigation application rate (1X), which could be accomplished by running the pivot at full travel speed without pulsing the nozzles. However, in the second section (S2) where the first three zones had sandy soil, the nozzles in Zones 1 through 3 would apply 2X irrigation rate by reducing the travel speed to 50% without pulsing the nozzles. But, to apply the 1X rate to the clay soil in Zone 4, the nozzles would be pulsed with 50% duty cycle. The irrigation strategy would also reduce the application rate to 50% (0.5X) for the boggy areas by pulsing the nozzles at 50% duty cycle. The system would also avoid irrigating the pond and turning areas by turning the pumps off (0X). It would also turn the end gun off in Sections 4 through 6 where a roadway passes outside the sections. For the variable rate irrigation, duty cycles for any management zones or forward speed could be set to any value between 0 and 100% in the control program.

Figure 9 shows schematic of the PLC program written to achieve the irrigation strategy illustrated inTable 1. The program uses a series of logic gates (NOT, AND, and OR gates) to control pumps and the drive system. A series of inverters (NOT gates) and four-input AND gates on the left analyzed three input switches to find out which of the eight sections the center pivot was in. Since the output of an AND gate was 1 (on) only when all inputs were 1, inverters were used whenever input values were 0 (off). For example, if all three switches were in the off position, the last AND gate would be turned on to signify that the pivot was in Section 0 (000). If the third switch was on, the output of the first AND gate would be on to signify that the pivot was in Section 1 (001).

Table 1. Example irrigation rates for the simulated conditions in Figure 7 (CW = clockwise; CCW = counter clockwise; X = irrigation rate for the clay soil).

Section	Micro-switch binary output	Travel speed & direction	Pump status, nozzle pulsing, relative irrigation rate				End
			Zone 1	Zone 2	Zone 3	Zone 4	Gun
S0	000	Full, CW	OFF, NO, 0.0X	OFF, NO, 0.0X	OFF, NO, 0.0X	OFF, NO, 0.0X	OFF
S1	001	Full, CW	On, No, 1X	On, No, 1X	On, No, 1X	On, No, 1X	ON
S2	010	50%, CW	On, No, 2X	On, No, 2X	On, No, 2X	On, 50%, 1X	ON
S3	011	50%, CW	On, 50%, 1X	On, No, 2X	On, 50%, 1X	On, 50%, 1X	OFF
S4	100	Full, CW	On, No, 1X	On, No, 1X	On, No, 1X	On, 50%, .5X	OFF
S5	101	Full, CW	On, No, 1X	On, No, 1X	On, No, 1X	OFF, NO, 0.0X	OFF
S6	110	Full, CW	On, No, 1X	On, No, 1X	On, 50%, .5X	On, 50%, .5X	ON
S7	111	Full, CCW	OFF, NO, 0.0X	OFF, NO, 0.0X	OFF, NO, 0.0X	OFF, NO, 0.0X	OFF

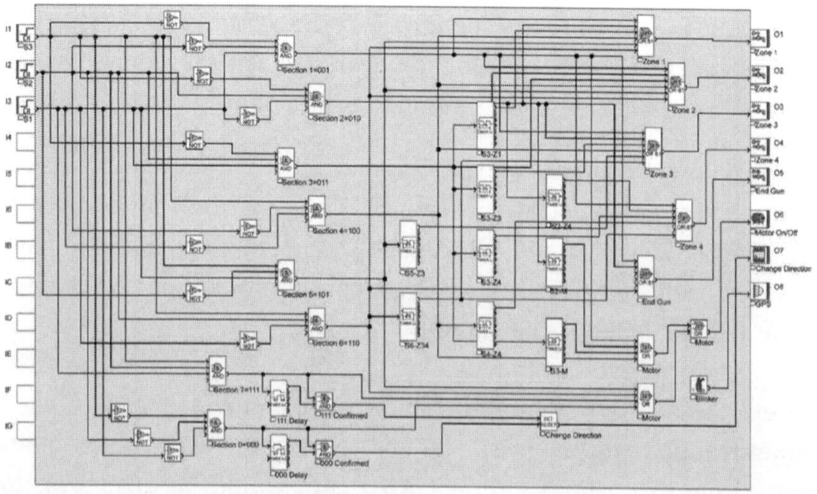

Figure 9. Schematic of the PLC program used to implement the irrigation strategy in Table 1.

If all three switches were on, the output of the AND gate for section 7 (111) would be on.

Five six-input OR gates on the rightmost side controlled the nozzles in four zones and the end gun. The six inputs on each OR gate were connected to position signals from Sections 1 through 6, respectively. All nozzles were off in Sections 0 and 7 which were used as a turning area. Since the output of an OR gate was 1 (on) if any of the inputs were 1, any position between Sections 1 and 6 could turn on the nozzles. If a nozzle was to be pulsed with a 50% duty cycle in a section, the input signal from that section would pass through a Li Timer with a 0.5 sec On-time and a 0.5 sec Off-time. The duty cycle of any nozzles could be changed by adjusting On-time and Off-time of the individual Li Timers. Missing inputs on any OR gates mean that the nozzle will not operate in the sections from where input connection was missing. In an actual VRI system, the nozzle pulsing time is normally one minute. However, due to the small size of the model and time limitation for demonstration, one second cycle was used instead.

The two four-input OR gates on the lower-right side controlled the motors in eight sections. The motors were mostly run at 100% speed except in Sections 2 and 3 (50% travel speed), where Li Timers were used to pulse the motors on or off with a 50% duty cycle. The duty cycle of the motors could be changed by adjusting On-time and Off-time of the Li Timers.

When the pivot reached Section 0 (leftmost) or Section 7 (rightmost), the position signal would set the Reversing Relay on or off to change the rotational direction of the ICPDemo to clockwise at Section 0 and counterclockwise at Section 7. There was a delay of 1.5 seconds before the motor changed its direction to confirm that the pivot actually was in Section 0 or Section 7. This time delay prevented the accidental direction change in any other sections due to a momentary false reading from the switches. This delay time interval could be changed during runtime from Parameters Menu on the PLC Screen. The LED simulating a GPS receiver was connected to the last output of the PLC and was programmed to blink at 1 Hz.

3. RESULTS AND DISCUSSION

A fully functional tabletop model of an intelligent center pivot was constructed according to the design specifications and was programmed using an example irrigation control strategy on a PLC. The system demonstrated varying application rates over sand, clay, a low-lying boggy area, a pond, and other features around the simulated field. The varying application rates were achieved by either pulsing the motors to reduce travel speed and apply a higher rate, pulsing the pumps to reduce the application rate, or turning the pumps off to avoid irrigation.

During the design and testing process, the durability of the pumps was found to be inadequate. The automobile windshield washer pumps used in this model were not designed to operate continuously, and could overheat if used continuously. This problem could be solved by replacing these pumps with models designed for continuous operation such as those normally used with small water fountains. However, due to short demonstration period, there was no need for continuous operation. Also, the two-planed drain system for the recirculation of water was not quite adequate, due to the surface tension of water clinging to the LEXANTM sheet on the top of the model. A greater slope was needed to achieve a full drain. For this model a squeegee was used to direct the residual water to the drain.

The final ICPDemo model has been successfully demonstrated in a number of presentations to various clienteles. This model could also be used as a teaching aid in a classroom for current and future students. It is our intention to continue to use this model in future irrigation extension activities such as field days, workshops, growers' production meetings, and in service training for county extension agents.

4. CONCLUSION

A tabletop intelligent center pivot demonstration model (ICPDemo) was designed and constructed to demonstrate and promote adoption of the innovative ICP technology. The final design was a small two-section center pivot covering a semi-circular field. The ICPDemo had four management zones plus an end gun whose application rates could be individually controlled over a number of control sections in the simulated field. The rotation speed of the pivot could also be controlled according to a simulated irrigation rate map. To demonstrate the variable nature of the ICP irrigation system, various soil types and field features were incorporated in the management zones, and the irrigation control strategy was programmed on a PLC. The ICPDemo is being used to showcase the ICP technology during a variety of irrigation extension activities.

ACKNOWLEDGEMENTS

This material is based upon work supported by NIFA/USDA, under project number SC-1700442, Technical Contribution No. 6390 of the Clemson University Experiment Station.

REFERENCES

1. Evans, G.W. and Harting, G.B. (1999) Precision Irrigation with Center Pivot Systems on Potatoes. Proceedings of ASCE 1999 International Water Resources Engineering Conference, Seattle, 8-11 August 1999, American Society of Civil Engineers, Reston.

2. Perry, C., Pocknee, S. and Hansen, O. (2003) A Variable Rate Pivot Irrigation Control System. In: Stafford, J. and Werner, A., Eds., ECPA 2003, Proceedings of the Fourth European Conference on Precision Agriculture, Wagner Academic Publishers, 539-544.

3. Khalilian, A., Henderson, W., Han, Y., Owino, T. and Niyazi, B. (2007) Scheduling Site-Specific Irrigation for Cotton Production Using a Linear Move System. Proceedings of the Beltwide Cotton Conferences, New Orleans, January 2007, National Cotton Council of America, Memphis, 883-888.

4. Khalilian, A., Han, Y.J. and Farahani. H.J. (2008) Site-Specific Irrigation Management. In: Eidson, G.W. and Sawyer, C.B., Eds., Proceedings of 2008 South Carolina Water Resources Conference, Charleston, 14-15 October 2008.

5. Han, Y.J., Khalilian, A., Owino, T.O., Farahani, H.J. and Moore, S. (2009) Development of Clemson Lateral Variable-Rate Irrigation System. Computers and Electronics in Agriculture, 68, 108-113.

6. Khalilian, A., Han, Y., Barnes, E.M. and Henderson, W. (2012) Water Use Efficiency of Different Cotton Cultivars.

Proceedings of the Beltwide Cotton Conferences, 3-6 January 2012, Orlando, National Cotton Council of America, Memphis, 517-521.

7. Khalilian, A., Qiao, X., Farahani, H., Han, Y. and Payero, J. (2013) Parameterization of the FAO AquaCrop Model for Irrigated Cotton in the Humid Southeast U.S.A. Proceedings of the Beltwide Cotton Conferences, San Antonio, 7-10 January 2013, National Cotton Council of America, Memphis, 522-527.

8. Perry, C. (2010) Personal Communication.

11

PRODUCTION RISK AND ADOPTION OF IRRIGATION TECHNOLOGY: EVIDENCE FROM SMALL-SCALE FARMERS IN CHILE

Ce´sar Salazar[1] • John Rand[2]

[1] Department of Economics and Finance and 'Ana´lisis Econo´mico Sectorial Aplicado' Research Group, University of Bio–Bio, Avenida collao 1202, Concepcio´n, Chile

[2] Department of Economics, University of Copenhagen, Copenhagen, Denmark

ABSTRACT

In most developing countries non-irrigation status often dominates adoption of traditional and modern irrigation technology. In this paper, we study the effect of production risk on irrigation technology choice among small-scale farmers in Chile, applying sample selection and discrete choice models. We find that more educated farmers, with credit access, receiving extension services, and living in communes with more adopters are more likely to use modern irrigation techniques. Moreover, production risk is often associated with adoption of traditional irrigation, and this risk often undermines a shift to more modern irrigation systems. Controlling for pre-conditions that determine irrigation choices clearly improves our understanding of small-scale farmer irrigation adoption decisions and we argue that weaker knowledge about and lower automatic diffusion of modern irrigation is a main obstacle for improving small-scale farmer productivity.

Keywords: Production risk Irrigation Technology adoption

1 INTRODUCTION

Since the green revolution, agriculture in developing countries has experienced a technological transformation process principally driven by an increasing demand for agricultural products. It has triggered a massive increase in the adoption of new technologies, such as the use of chemical inputs, pesticides, traditional irrigation, and high-yielding seed varieties, with significant benefits in terms of productivity and poverty reduction (Evenson and Gollin2003). However, concerns have lately been raised on the grounds of sustainability, as the intensive use has consequences for environment and human health (Ruttan 2002).

The adoption of traditional irrigation methods has shown to be beneficial to reduce farmers' vulnerability to weather conditions, making production and incomes more stable. However, traditional irrigation has also increased the pressure on water reservoirs and created pollution problems.[1]

The aim of this paper is to study the effect of production risk on irrigation technology choices and discuss adoption of modern irrigation in settings where most rural households depend exclusively on rainfall for water supply. This is a particular feature in developing countries and zones characterized by abundance of precipitation.[2] We argue that the latter may give rise to potential sample selection problems in irrigation choices, which needs to be considered when analyzing the association between production risk and farm technology adoption. For modeling these irrigation choices, we as an example use agricultural census data of potato family farmers in Chile.

Chile has the challenge to transform the Chilean agricultural sector into a world-wide agricultural production and food powerhouse (MINAGRI 2006). However, such a transformation process requires incorporating small holder family farmers into productive agricultural units not only serving themselves (home-production) but as integrated highly productive active agents contributing to the achievement of this objective. The Agricultural Development Institute (ADI) has continually provided technical support to the family agriculture in areas such as investment in human and productive capital, credit provision, access to markets, recuperation of eroded soil and irrigation. In spite of this effort and progress observed in the last decade, the technological gap between large and small-scale farmers persists. For example, 89 % of total land under modern irrigation systems is held by larger farmers (FAO 2009). This fact is what motivates this study to analyze the factors explaining the slow adoption process among small-scale farmers, where the main focus is on the importance of differences in production risk faced by small versus large farmers.

The association between production risk and adoption of modern irrigation is quite straight forward. Risk-averse farmers may not adopt modern irrigation if it raises production risk by increasing yield variability too much. In contrast, if the new technology has the potential to attenuate variability in economic outcomes, then it would be more rapidly adopted among risk-averse small scale farmers. Even though literature on adoption of modern irrigation technology is well-documented, there is little evidence concerning production risk and irrigation choices. Two exceptions are Koundouri et al. (2006) and Torkamani and Shajari (2008). Both studies find support of a positive and significant effect of production risk on adoption decisions, arguing that new irrigation technology is a risk-decreasing input. However, both studies were carried out in settings where the knowledge of irrigation benefits is well-known and the new technology seems to show some degree of consolidation, which only allow to model changes from traditional to modern irrigation systems.

In Chile the pre-conditions are different. Due to abundance of precipitation mainly in the southern territories in Chile, around 83 % of potato farmers do not use irrigation, which amounts to 66 % of registered farming land. Exploiting these differences in natural pre-conditions, we follow Foudi and Erdlenbruch (2012) applying a sample selection approach to study the risk-decreasing properties of irrigation under joint estimation of production and production risk. However, to our knowledge, no previous studies have extended this approach to the decision of shifting from conventional farming to modern irrigation technologies.

Moreover, in light of the high drought risk observed in Chile understanding the underlying mechanisms of irrigation adoption is of fundamental value. Historically droughts are frequent and affect 25 % of the continental territory. Furthermore, climate change studies estimate an intensification of aridity in the north zone, advances of the desert toward the south, and less precipitation in the central zone at the end of the 21st century (FAO 2010). Thus, the augment in temperature and decline in precipitation along with an increasing demand for water resources foresee a higher drought risk for the next decades. Therefore, improvements in water efficiency by the adoption of modern technology are likely to become essential to reduce the vulnerability of agriculture in the future.

The rest of the article is organized as follows: Sect. 2 reviews the main arguments enhanced by literature as determinants of technology adoption. Section 3 describes the empirical strategy and data used to carry out this study. Section 4 presents the main results, and Sect. 5 concludes the article.

2 CONCEPTUAL FRAMEWORK ON ADOPTION OF NEW TECHNOLOGY

There is an extensive literature aimed at explaining the process of technology adoption.[3] Lower adoption rates are generally associated with imperfections in credit markets, information, agro-ecological characteristics, input and output markets as well as inadequate incentives associated with farm tenure arrangements, problems mainly observed in developing countries (Feder et al.1985; Bardhan and Udry 1999).

A key element relates to the reduced credit opportunities farmers face. Lack of access to credit in rural areas arises as a consequence of asymmetric information and enforcement problems in the lender-borrower relation. This is particularly augmented among small farmers in developing countries due to the lack of sufficient collateral (Udry1994). Thus, when innovations require significant levels of investment, credit constraints may turn out to be a limiting factor in adoption decisions. In addition, as the diffusion of innovation is scarce and fails to reach potential users, farmers may face trouble in obtaining the maximum benefit from shifting to modern technologies. The optimal use of new production methods involves adoption of new technologies to the local conditions, which implies a costly learning process (Foster and Rosenzweig 1995). Imperfect knowledge about the usage of new technology constitutes a barrier that delays the adoption process. Furthermore, it is expected that new technologies are more likely to be adopted in rural zones with better support infrastructure and localized near main commercial centers, in which the availability of both complementary inputs and maintenance services are more abundant (Sunding and Zilberman 2001). In addition, tenurial arrangements that characterize the landlord-tenant relationship and land rent contracts can affect adoption decisions. For instance, small farmers without land of their own and facing short-term land contracts are less probable to adopt innovations since they may be unable to enjoy the long-term benefits of doing so (Bahduri 1973). Finally, agro-ecological characteristics that capture environmental heterogeneity are essential to attain a better understanding of the process of adopting sustainable agricultural technologies (Dinar and Yaron 1990; Lee 2005). In particular, modern irrigation methods have a higher relative advantage in zones with lower land quality which lower water-holding capacity (Caswell and Zilberman 1986; Dinar and Zilberman 1991).

In relation to household characteristics, variables that control for differences in farm size, off-farm opportunities, human capital, and gender play an important role to capture the intra-household dynamic in explaining adoption levels (Doss 2006). One key factor less explored in literature is the role of risk in adopting new technology (Foster and Rosenzweig 2010). Rural households must deal with high-income fluctuations as exposed to a variety of uncontrollable factors such as climate conditions, economic fluctuations, policy uncertainty, and individual-specific shock[4] (Bardhan and Udry 1999; Dercon 2002). In this context, decision making under uncertainty is characterized by risk because some possible outcomes have an undesired effect. If we assume farmers are risk averse, they would adjust their behavior implementing actions aimed at smoothing income and/or consumption (Dercon 2002). However, when credit and insurance markets are absent or incomplete, poor farmers characterized by limited landholdings, few assets, and lower schooling face serious constraints to smooth consumption.[5] Consequently, it is expected that farmers depend on ex-ante management strategies to smooth income, which would imply preferences for traditional technology with lower expected returns. It may be true for new irrigation methods that due to higher exigencies of knowledge in its use, adoption would raise the likelihood of failure in spite of higher expected benefits. However, modern irrigation technology is also recognized for being a risk-decreasing input when reducing dependence on rainfall and water availability, which make yields more stable. Koundouri et al. (2006) and Torkamani and Shajari (2008) found a positive effect of production risk on switching from traditional to modern irrigation technology, supporting a risk-decreasing effect.

3 AGRICULTURAL SECTOR AND FAMILY FARMING IN CHILE

In this paper, we examine the effect of production risk on irrigation choices, focusing on a sample of family farmers in Chile among which modern irrigation is an incipient agricultural practice. The agricultural sector is important in Chile, accounting for 3.5 % of the Gross Domestic Product (GDP) and employing around 12 % of the country's labor force. More than 50 % of production is commercialized in international markets, which makes the sector highly dependent on world market prices. The sector benefits from huge climate diversity, which allows an ample range of cropping activities. This agro-ecological diversity is considered one of the Chile's main comparative advantages. Some major agriculture products consist of fruits, including grapes, apples, pears, peaches, and berries; horticulture, including garlic, onions, and asparagus; and cereal and tuber such as wheat, maize, and potatoes.

Family agricultural production represents one-third of Agricultural Gross Domestic Product (AGDP) and contributes with 1.2 % of the GDP in Chile. In addition, this segment controls 85 % of the farms in the country, and generates 60,000 direct and indirect jobs, benefiting 1.2 million of people. Furthermore, family agricultural producers are one of the main food suppliers in the country, providing around 60 % of food consumed in the domestic markets (INDAP2011). Family farmers are characterized by being mainly involved in traditional farming activities such as annual crops, horticulture, and extensive cattle farming. They operate small plots at low levels of working capital and often face credit constraints. For simplification, the Ministry of Agriculture defines family farmers as those holding 12 hectares or less of basic irrigation (HBI) (FAO 2009). We follow this definition in this paper.

4 DATA

This study uses data from the VII National Agriculture and Forestry Census carried out in Chile (INE 2007).[6] All data is at the household level, except for tenancy status, which was collected at the plot level. Household level data is combined with location level data for soil and climate characteristics.

The Census only reports information on output and input variables for annual crops. Farmers involved in other economically relevant cropping activities, such as fruit trees and horticulture, were unfortunately not asked to report production and input volumes. Given that our framework relies on the estimation of the production function to compute risk measures, our data is not suitable to investigate the association between irrigation technology and production risk to a larger segment of Chile's agricultural sector, other than farms growing annual crops. In spite of the limitations outlined above, census data allow us to deal with a recurring problem observed in site-specific data, in which a few villages or areas are surveyed, and data lack sufficient variation capturing the agricultural potential (Doss 2006). The Census gives us an exceptional opportunity to exploit the climate diversity that characterizes the Chilean territory. Due to its unusual length and the diversity of its landscapes and ecosystems, Chile has many distinct climates, ranging from desert to frosty polar in the extreme south. A national coverage of the Census running more than 4000 km from north to south guarantees a unique setting to study the role of natural pre-conditions (rainfall and soil characteristics) in the identification of selection into irrigation.

We choose to focus on potato farming. The choice of this crop relies on the economic and social importance for agricultural income-dependent farmers. This crop is the third more relevant annual crop in term of surface and occupies the second position when production value is taken into account. Apart from its economic relevance, potato farming is important from a social perspective due to the fact that it demands great amounts of labor mainly during harvest season, which is particularly transcendental in those zones with higher unemployment. Moreover, this crop extends across the whole territory and its distribution can clearly be organized in agro-climatic zones well differentiated (INIA 2006). This feature allows us to pick up agricultural potential with more precision. Another issue is that we are not able to distinguish amounts of inputs per crop. Thus, to ensure a clean association between observed inputs and potato production, we decided to focus on the sample of potato farmers whose entire land is allocated to this crop. This resulted in a sample of 7274 observations.

Farmers report number of hectares irrigated by different irrigation systems. We aggregate this information into three irrigation statuses: non-irrigation, traditional irrigation and modern irrigation.[7] Table 1 shows the number of farmers and hectares that fall into each category of irrigation.

Table 1. Number of farmers and hectares per irrigation category

Category	N°	%[a]	Hectares	% Hectares
No-irrigation	6083	83.0	4094	65.6
Irrigation	1245	17.0	2146.1	34.4
Traditional	1175	94.4	1852.5	86.3
Flood	417	35.5	692	37.4
Furrow	831	70.7	1139	61.5
Other	17	1.4	21.5	1.2
Modern	70	5.6	293.6	13.7
Sprinkler	53	75.7	215	73.2
Center pivot	2	2.9	64	21.8
Drip	8	11.4	6.6	2.2
Micro sprinkler	7	10.0	8	2.7

Source: Own elaboration based on Censual data, 2007

[a]Traditional irrigation does not add up 100 % because some farmers adopt more than one irrigation system. Moreover, some farmers use both traditional and modern irrigation. In this case, they are defined as modern irrigators

The figures show that 17 % of farmers (and 34.4 % of farmed land) use some system of irrigation. In addition, modern irrigation is adopted by 5.6 % of irrigators, covering 13.7 % of the total irrigated hectares. Regarding traditional irrigation, farmers using furrow technology amount to 70.7 % of the total (and 61.5 % as measured by irrigated hectares). On the other hand, sprinkler technology seems to be the more broadly used system among modern irrigators since three out of four adopt this method. Table 2 depicts the summary statistics for inputs, environmental, and location variables by irrigation category. Table 3 shows descriptive statistics for farmers' characteristics and institutional aspects by irrigation category.

Table 2. Mean and standard deviation for inputs, environmental and location variables per irrigation category

Category	Modern irrigation		Traditional irrigation		Non-irrigators		Whole sample	
	Mean	Std Dev	Mean	Std Dev	Mean	Std Dev	Mean	Std Dev
Production (kilos)	114,114.30	351,117.10	18,721.30	41,402.10	9033.70	48,555.60	11,739.50	61,987.70
Yield (kilos/hectares)	13,636.14	10,693.16	11,791.30	8734.11	11,227.95	8531.67	11,341.29	8591.53
Total surface (hectares)	24.93	36.70	4.44	12.73	15.81	25.25	14.08	24.21
Cultivated land	4.76	9.94	1.47	3.09	0.66	1.89	0.83	2.54
Total labor (N°)	6.14	7.30	4.18	6.76	3.46	4.57	3.59	5.01
Capital (Chilean pesos[a])	6,862,143	8,778,120	3,394,974	7,209,798	2,408,747	4,253,057	2,608,270	4,932,854
Fertilizer machine use								
No use	84 %		91 %		97 %		96 %	
Draft animal	0		0.05 %		0.03 %		0.04 %	
Draft mechanical	15 %		7 %		2.4 %		3.3 %	
Rainfall (mm)	963.1	513.8	467.3	343.7	1460.7	473	1319.5	405.9
Standard deviation	208.7	100.8	189.9	90.7	220.3	30.4	215.3	48.1
Skewness	−0.27	0.858	−0.043	0.818	0.033	0.794	0.018	0.79
Soil quality (ratio)	43.0 %	0.26	43.0 %	0.23	44.0 %	0.24	44.0 %	0.24
Location variables								
North zone	17.0 %		22.0 %		0.03 %		3.0 %	
Central zone	64.0 %		78.0 %		40.0 %		47.0 %	
South zone	17.0 %		0.0 %		59.0 %		49.0 %	
Observations	70		1153		6051		7274	

1 US$ corresponds to approximately 500 Chilean pesos

Source: Own elaboration based on Censual data, 2007

Table 3. Farmers' and institutional characteristics per irrigation category

Category	Modern irrigation		Traditional irrigation		Non-irrigators		Whole simple	
	N°	Mean	N°	Mean	N°	Mean	N°	Mean
Age	–	54.44	–	55.03	–	56.36	–	56.13
Male	55	79.0 %	876	76.0 %	4236	70.0 %	5165	71.0 %
Reside in plot	54	77.0 %	657	57.0 %	4780	79.0 %	5528	76.0 %
Education								
No education	2	2.9 %	67	5.80 %	163	2.70 %	233	3.20 %
Incomplete elementary	29	41.4 %	548	47.50 %	3685	60.90 %	4263	58.60 %
Complete elementary	8	11.4 %	209	18.10 %	938	15.50 %	1157	15.90 %
Incomplete secondary	9	12.9 %	90	7.80 %	411	6.80 %	516	7.10 %
Complete secondary	6	8.6 %	122	10.60 %	381	6.30 %	509	7 %
Incomplete technical	0	0.0 %	6	0.50 %	24	0.40 %	29	0.40 %
Complete technical	11	15.7 %	55	4.80 %	248	4.10 %	313	4.30 %
Incomplete college	1	1.4 %	12	1 %	48	0.80 %	58	0.80 %
Complete college	4	5.7 %	44	3.80 %	151	2.50 %	196	2.70 %
Dependence								
>75	25	35.7 %	205	17.80 %	1355	22.40 %	1586	21.80 %
>50 and <75	12	17.1 %	98	8.50 %	714	11.80 %	829	11.40 %
>25 and <50	13	18.6 %	186	16.10 %	1222	20.20 %	1426	19.60 %
<25	20	28.6 %	664	57.60 %	2753	45.50 %	3433	47.20 %
Credit	43	61.40 %	224	19.40 %	1767	29.20 %	2073	28.50 %
Secure tenure (hectares)								
Own with registered title	1391.5	73.6 %	24034.8	67.6 %	68,787.3	31.8 %	94,213.6	37.1 %
Rented	305.5	16.2 %	7625.5	21.4 %	6004.6	2.8 %	13,935.6	5.5 %
Occupied	18.1	1.0 %	10.5	0.0 %	64,069.9	29.6 %	64,098.5	25.2 %
Other	174.6	9.20 %	3884.4	11.00 %	77,717.9	35.90 %	81,776.9	32.20 %
Extension services	27	38.60 %	92	8 %	974	16.10 %	1091	15 %
Participation in organizations	22	31.40 %	351	30.40 %	914	15.10 %	1287	17.70 %
Average number per locality	–	76	–	321	–	546	–	945
Observations	70		1153		6051		7274	

Source: Own elaboration based on Censual data, 2007

Production yield increases as irrigation becomes more sophisticated. The same is true for cultivated area, indicating that irrigation is more likely to be adopted by large farmers. Irrigators mostly reside in zones with lower levels of precipitation. In particular, both modern and traditional irrigators predominate in the central zone. Conversely, non-irrigators are mainly located in southern locations, which experience the highest levels of rainfall.

Education seems to be important to explain adoption of modern technology. Furthermore, modern irrigators seem more agricultural income-dependent than farmers using traditional irrigation. In addition, only 28.5 % of farmers report to have used credit instruments. In relation to extension services, 38.6 % of farmers with improved irrigation methods state to have received extension services in the last 2 years, which compares to the 8 and 16 % observed among traditional irrigators and non-irrigators, respectively. There is a clear trend for irrigators to participate in organizations. Whereas participation reaches around 30 % among irrigated farms, this figure is only 15 % among non-irrigators. Regarding land property status, owning land with a registered title and rented land is more likely among irrigators. While this amount is around 90 % of total land among irrigated farms, it constitutes only 34.6 % of land held by non-irrigators.

Table 4 shows the number and percentage of farmers per water source and irrigation category, and Table 5explores changes in water rights uses across water sources and type of irrigation, respectively.

Up to 44 % of non-irrigators report to have wells as water source in the plot. Differences in costs associated with water extraction between surface and ground water sources naturally impose a restriction to adopt irrigation. However, these restrictions seem to be less important for the adoption of modern irrigation. Wasting water becomes much more costly when pumped from ground water sources[8] and the availability of water sources other than rivers may promote adoption of new technology (Caswell and Zilberman 1986).

Table 4. Number and percentage of farmers per water source and irrigation category

Water Source	Modern irrigation		Traditional irrigation		Non-irrigators		Whole simple	
	N°	%[a]	N°	%[a]	N°	%[a]	N°	%[a]
Well	17	24.3	105	8.9	57	44.9	179	13.0
Spring	16	22.9	118	10.0	33	26.0	167	12.2
River	25	35.7	625	53.2	19	15.0	669	48.8
Stream	8	11.4	184	15.7	30	23.6	222	16.2
Seasonal regulation reservoir	1	1.4	19	1.6	0	0.0	20	1.5
Yearly regulation reservoir	2	2.9	62	5.3	0	0.0	64	4.7
Lake	0	0.0	1	0.1	0	0.0	1	0.1
Lagoon	1	1.4	2	0.2	0	0.0	3	0.2
Other	5	7.1	93	7.9	8	6.3	106	7.7

Source: Own elaboration based on Censual data, 2007

[a] It does not add up 100 % because some farmers have more than one water source

Table 5. ater rights status per irrigation category

Water source	Registered in Real State Registrar or in the public land registry of the General Bureau of water			Rented			Title in process of regularization			Use de facto		
	Modern (%)	Traditional (%)	Non-irrigated (%)	Modern (%)	Traditional (%)	Non-irrigated (%)	Modern (%)	Traditional (%)	Non-irrigated (%)	Modern (%)	Traditional (%)	Non-irrigated (%)
Well	47.1	38.1	12.3	0.0	2.9	0.0	5.9	10.5	14.0	47.1	42.9	71.9
Spring	12.5	39.8	3.0	0.0	0.8	0.0	6.3	2.5	0.0	81.3	55.1	93.9
River	64.0	82.7	21.1	0.0	5.9	0.0	4.0	1.0	0.0	32.0	10.1	73.7
Stream	50.0	40.8	10.0	0.0	1.6	3.3	25.0	1.6	20.0	25.0	54.3	60.0
Seasonal reservoir	100.0	57.9	0.0	0.0	21.1	0.0	0.0	0.0	0.0	0.0	21.1	0.0
Yearly reservoir	50.0	75.8	0.0	50.0	8.1	0.0	0.0	0.0	0.0	0.0	14.5	0.0
Lake	0.0	100.0	0.0	0.0	0.0	0.0	0.0	0.0	0.0	0.0	0.0	0.0
Lagoon	100.0	100.0	0.0	0.0	0.0	0.0	0.0	0.0	0.0	0.0	0.0	0.0
Other	20.0	45.1	0.0	0.0	4.3	0.0	0.0	4.3	0.0	60.0	35.5	75.0
Total	45.4	64.7	10.2	1.3	4.7	0.7	6.7	2.2	9.5	45.3	26.4	74.8

Source: Own elaboration based on Censual data, 2007

Around 85 % of water sources held by non-irrigators fall into categories classified as insecure tenancy—title is in process of regularization or use de facto—in contrast to that observed in both modern and traditional irrigation. Undoubtedly, usage rights are crucial to create correct incentives to undergo innovations that require time to become profitable.

5 ESTIMATION PROCEDURE

The procedure to analyze the effect of production risk on adoption decisions follows two steps. In the first stage, we estimate the moments of a production function to proxy production risk. In the second stage, the estimated moments are incorporated to explain adoption of modern irrigation in discrete choice models.

5.1 Production risk

The moments of the production distribution are estimated by following a sequential procedure in which production is regressed against a set of inputs (Antle 1983; Antle 1987).[9] The model is specified as follows.[10]

$$y_i = f(x_i, z_i, h_j; \beta) + \varepsilon_i, \tag{1}$$

where $i = 1 \ldots N$ denotes individual farmers, y_i is the logarithm of output (potatoes) measured in kilos, x_i is a vector of conventional inputs including land, labor, capital, and fertilizer. Land is measured in hectares,[11] labor is the sum of both family and hired labor, and capital is the value of physical assets.[12] All these variables are expressed in logarithms. We proxy the amount of fertilizer by including a categorical variable that takes the value zero (0) if the farmer does not use any fertilizer machine, one (1) if the farmer uses a draft animal fertilizer machine and two (2) if the farmer uses a draft mechanical fertilizer machine. Z_i is a vector of farmers' characteristics including age (expressed in logarithm), education[13] and agricultural dependence.[14] Additionally, we include variables (h_j) to capture variations in soil quality and rainfall[15] (expressed in logarithm) in locality j as well as regional differences. Soil quality is measured by the percentage of non-eroded and slightly eroded soil[16] and rainfall was constructed as the cumulative precipitations in the season 2010–2011. Unobserved regional differences are picked up by including indicator variables for each region. ε_i is the identically and independently distributed (iid) error term.

Under expected profit maximization, explanatory variables are assumed to be exogenous, and thereby ordinary least squares (OLS) of (1) produce consistent and efficient estimates of the parameter vector β (Koundouri et al.2006). The s central moment of production conditioned on inputs about its mean is defined as:

$$\mu_s(.) = E\{[y(.) - \mu_1]^s\}, \tag{2}$$

where μ_1 denotes the mean of production. Thereby, the estimated errors ε^i=yi−f(xi,zi,hi;β)ε^i=yi−f(xi,zi,hi;β) from (1) are estimates of the first moment of the production distribution. To compute estimates of the second moment, the estimated errors ε^ε^ are squared and regressed on the same set of inputs as in (1):

$$\hat{\varepsilon}_i^2 = g(x_i, z_i, h_i; \delta) + \bar{\varepsilon}_i. \tag{3}$$

OLS provides consistent estimates of the parameter δ, and the predicted values are consistent estimates of the second moment of production distribution (Antle 1983). Estimation of the third moment follows the same procedure. This approach exploits the use of cross-sectional data assuming that the moments vary among farmers depending on input level, farms' characteristics and environmental conditions. Conditional moments are thereby used as measures of volatility to rank farmers in terms of production risk.

5.2 Irrigation choices

In the second stage, the estimated moments are incorporated as explanatory variables in a discrete choice model. Additional control variables include a dummy denoting if the farmer is male; farmers' age measured in years (in log); farmers' level of education; a categorical variable that captures the degree of dependence on agricultural activity; a dummy variable indicating if the farmer lives in the plot; the farm's size measured in total hectares (expressed in a logarithm); capital value of agricultural machinery and tools (expressed in a logarithm); secure tenure measured by the ratio between the sum of family-own hectares and rental land over total hectares; a dummy variable indicating if the farmer had access to credit during the last 2 years; a dummy variable indicating participation in any agricultural organization; a dummy variable denoting if the farmer received extension services during the last 2 years; percentage of non-eroded and slightly eroded soil; number of both modern and traditional adopters per locality regardless of crops; and a set of dummy variables to control for unobserved spatial differences in technology choice across zones.

We propose three alternative models. First, we model irrigation decisions as an ordered choice such that each category corresponds to a superior level of irrigation technology. One may expect a monotonic relationship between risk production measures and specific irrigation technologies as the capability of irrigation methods to reduce production risk increases with the level of technicality. Second, we assume that irrigation choices are unordered at different production risk levels. Despite reduction in production risk from shifting from a non-irrigation status to a traditional irrigator seems quite evident, gains from adopting modern irrigation for a former traditional irrigator are more unclear as new technology involves uncertainty. Finally, we discuss potential selection problems when analyzing the shift from traditional to modern irrigation. We argue that the choice and economic benefits of irrigation adoption depend on natural pre-conditions to irrigate. Thus, when considering only irrigation data, we might lose track of some people who are eligible to adopt modern irrigation. If this characteristic is systematic, our standard probit estimation may lead to inconsistent estimates (Wooldridge 2010). We, therefore, also estimate a binary response model with sample selection (Van de Ven et al. 1981) to address this problem. Denoting Y as the irrigation decision observed with the value of 1 if the farmer irrigates and 0 otherwise, the selection equation can be expressed as follows:

$$\Pr[Y_i = 1] = \Pr\left[Y_i^* > 0\right] = \Pr\left[w'\varphi + \xi_i > 0\right] = \Pr\left[\xi_i < w_2'\varphi\right], \quad (4)$$

where Y∗iYi∗ corresponds to a latent variable that depends on a set of variables *w*, among which we include those to allow for identification of the vector φφ. Basically, we argue that selection into irrigation mainly lies in the fact that despite the presence of water resources in the farm, a significant number of farmers do not make use of these water courses for irrigation. Tables 4 and 5 show that the type of water source and water rights statuses may play a role here. Thus, we assume that irrigation decisions respond to the availability of water sources in the farm and moments of rainfall distribution. We account for water sources by defining a dummy variable that takes the value of 1 if the farm holds at least one source for irrigation.

6 RESULTS

Estimation of the selection equation is depicted in Table 6. As expected, accessibility to water sources plays a crucial role in explaining the shift from complete reliance on rainfall to adoption of irrigation systems. Furthermore, farmers who reside in locations with lower levels of precipitation, higher variance, and skewness are more likely to irrigate.[17] Similar results were found in related literature before. For example, Negri et al. (2005) examines the independent effects of climatic mean and variance on the probability of adopting irrigation in USA. The authors find that higher temperatures and less rainfall increase irrigation, and most importantly that the tails of precipitation distributions and water availability are primary determinants of presence of irrigation. Furthermore, Foudi and Erdlenbruch (2012) study the role of irrigation as a self-insurance instrument in the management of production risk in France. They find that irrigation decision depends on the decision to purchase yield insurance as well as the mean and variance of water availability. We compute a likelihood ratio test assuming under the null hypothesis$\rho = 0$, that is, the log-likelihood for the probit model with the sample selection is equal to the sum of the log-likelihoods of estimating a probit model for modern irrigation and the selection equation separately. The evidence rejects the null hypothesis suggesting that farmers irrigating may not be a random sub-sample of total rural households, which supports the use of data on non-irrigators to correct for sample selection.

Table 6. Estimates of the selection equation

Variables	Probit with sample selection Irrigation = 1
Water source	4.63 (0.39)***
Rainfall	–0.95 (0.16)***
Standard deviation of rainfall	0.005 (0.001)***
Skewness of rainfall	0.31 (0.12)**
Athrho	1.13 (0.39)**
LR test ($\rho = 0$)	13.21**
Observations	7274

Standard errors in parentheses

Independent variables of the selection equation are not shown due to space reasons

*** $p < 0.01$, ** $p < 0.05$, * $p < 0.1$

We employ a Translog functional form to estimate the parameters of the production function. Results are not shown but can be obtained from the authors upon request. We use these estimates to compute the moments of the production function, which are our proxies for production risk in the irrigation choice models. Table 7 reports the estimated coefficients and standard errors in parenthesis for the irrigation choice models.[18] The first column shows the estimated coefficients that result from modeling irrigation as a binary variable. Column 2 depicts the estimates for the ordered probit model, which assumes an order in irrigation choice. The third column shows the estimated parameters for the multinomial probit model. We assume non-irrigation status as the baseline. Column 4 shows the estimated coefficients for the model of modern irrigation for the selected sample of irrigators. In this case, non-irrigators were dropped from the sample. Column 5 presents the results for the model of adoption of new technology conditioned on whether a farmer is already an irrigator. Regardless of the approach used, education, capital, extension, credit access, number of adopters, and rainfall are statistically significant to explain adoption of irrigation in general, and modern irrigation in particular.[19] Education, capital, credit, and extension services are positively associated with adoption. Similar results have been found in developing countries in the literature before. For example, He et al. (2007) investigate the determinants of farmers' decisions to adopt rainwater harvesting and supplementary irrigation technology in China, finding that socioeconomic characteristics such as age and education, and institutional factors associated with extension, assistance, training, and credit services affect positively irrigation adoption decisions. Household resource endowment variables, showing the ability of farmers to operate their land, were also found significant in explaining farm irrigation decisions in Ghana (Dessalegn 2005). Results confirm the role of education as a complement to adoption when it is knowledge intensive, and benefits are linked to farmers' management capacity. Furthermore, they enhance the importance of access to credit in decisions that involve financing needs. Finally, the implementation of water-saving technologies demands more knowledge about environmental factors, i.e. evaporation conditions, to reach efficiency in water application, which makes extension services from government agencies be more crucial.

Table 7. Estimates of the models of technology adoption

Variables	Probit Irrigation = 1	Ordered probit+ Level: 1, 2, 3	Multinomial probit		Probit on selected sample Modern irrigation = 1	Probit with sample selection Modern irrigation = 1
			Modern	Traditional		
Moments of production distribution						
Mean	0.036 (0.029)	0.018 (0.018)	0.053 (0.077)	0.050 (0.046)	0.135 (0.100)	0.1498 (0.092)
St. Dev.	0.521 (0.143)***	0.154 (0.125)	−0.325 (0.351)	0.844 (0.195)***	−1.129 (0.292)***	−0.943 (0.271)***
Skewness	0.039 (0.017)**	0.005 (0.016)	−0.013 (0.040)	0.068 (0.024)**	−0.108 (0.034)***	−0.087 (0.032)***
Household characteristics						
Gender	0.133 (0.073)*	0.129 (0.055)**	0.055 (0.192)	0.224 (0.108)**	−0.261 (0.233)	−0.235 (0.223)
Age	−0.164 (0.105)	−0.117 (0.084)	0.209 (0.294)	−0.321 (0.159)**	0.192 (0.348)	0.139 (0.324)
Education	0.068 (0.016)***	0.062 (0.013)***	0.144 (0.036)***	0.071 (0.028)**	0.131 (0.0472)***	0.125 (0.045)***
Reside in farm	0.123 (0.072)*	0.053 (0.0532)	0.399 (0.227)*	0.128 (0.108)	0.695 (0.241)***	0.647 (0.239)***
Dependence	0.017 (0.029)	−0.005 (0.024)	0.017 (0.070)	0.028 (0.043)	−0.016 (0.080)	−0.0006 (0.0751)
Capital	0.009 (0.004)**	0.008 (0.003)**	0.025 (0.012)*	0.009 (0.006)	0.029 (0.013)**	0.0254 (0.012)**
Land	0.138 (0.036)***	0.085 (0.035)**	0.366 (0.099)***	0.121 (0.050)**	0.117 (0.089)	0.121 (0.082)
Institutional aspects						
Secure tenure	0.223 (0.095)**	0.239 (0.068)***	0.650 (16.302)	0.261 (0.131)**	0.601 (0.470)	0.497 (7.043)
Extension	0.235 (0.098)**	0.260 (0.088)***	0.639 (0.184)***	0.112 (0.158)	0.481 (0.238)**	0.434 (0.224)*
Credit	0.161 (0.087)*	0.213 (0.071)***	0.638 (0.181)***	0.006 (0.136)	0.457 (0.22)**	0.475 (0.204)**
Participation	0.205 (0.070)**	0.098 (0.054)*	0.056 (0.180)	0.355 (0.105)***	−0.213 (0.200)	−0.136 (0.182)
N° irrigat (T)	0.0019 (0.0001)***	0.0012 (0.000)***	0.001 (0.000)***	0.002 (0.0002)***	−0.001 (0.0002)***	−0.0011 (0.0002)***
N° irrigat (*M*)	−0.0007 (0.0006)	−0.001 (0.0003)***	0.0029 (0.001)**	−0.002 (0.0008)**	0.002 (0.0009)***	0.002 (0.0009)**
Environmental factors						
Soil quality	−0.652 (0.154)***	−0.138 (0.117)	−0.813 (0.457)*	−0.924 (0.226)***	0.661 (0.454)	0.586 (0.480)
Rainfall	−1.191 (0.067)***	−0.885 (0.052)***	−0.383 (0.237)	−1.791 (0.094)***		
North zone	1.151 (0.300)***	0.375 (0.098)	2.249 (5.865)	1.599 (5.867)	−0.524 (0.310)*	−0.404 (0.295)
South zone	−0.530 (0.131)***	−0.859 (0.131)	−0.794 (0.250)***	−28.95 (3.388)***		
Constant	6.435 (0.693)***		−2.555 (16.397)	10.117 (1.027)***	−2.873 (1.430)**	−2.823 (7.087)
Cut1		−4.636 (0.530)***				
Cut2		−2.246 (0.495)				
Observations	7274	7274	7274		1223	1223

Bootstrapped standard errors in parentheses, 1000 replications

*** $p < 0.01$, ** $p < 0.05$, * $p < 0.1$

Furthermore, the number of adopters of both traditional and modern irrigation at the local level is also significant. This result is in line with the literature addressing the role of social learning spillover on adoption of new technology among farmers that belong to related networks or adjacent geographical areas. For example, Foster and Rosenzweig (1995) find that neighbors' experience influence rates of adoption and profitability of high-yielding varieties (HYV) in India. The later suggests that farmers involved in a bigger network of experienced farmers with new technology are more likely to adopt the new technology and are relatively more profitable. Furthermore, Conley and Udry (2010) investigate the role of social learning in the diffusion of a new agricultural technology in Ghana. They find support of farmers adjusting their inputs in accordance with those of their neighbors who were surprisingly successful in previous periods. Finally, Bandiera and Rasul (2006) present evidence on how farmers' decisions to adopt a new crop relate to the adoption choices of their network of family and friends in Mozambique. The authors find farmers are more likely to adopt when other farmers in their network also adopt.

When looking at the results from the probit model and ordered probit model (column 1 and column 2, respectively), we also find significant effects of gender, farm size, secure tenure, and participation in organization. Gender differences in technology adoption are discussed in Ndiritu et al. (2014). The authors find a gender gap in the adoption pattern for some farm technologies in Kenya, indicating the existence of certain socioeconomic inequalities and barriers for female farmers. Furthermore, the larger the farm is the more probable adoption of irrigation technology. Just et al. (1980) show that given the information and transactions costs associated with innovations, as these costs increase, the critical size that make the adoption profitable also increases. The latter implies that new technology, which involves larger information/transactions costs are less likely to be adopted by smaller farmers. Empirical works confirm this prediction (Negri and Brooks 1990; Shrestha and Gopalakrishnan 1993; Skaggs 2001; Daberkow and Mcbride 2003; Schuck et al. 2005).

Farmers that hold a larger ratio of both owned land and land under property rental contracts are more likely to irrigate. Similar results are found in the literature before. For example, Soule et al. (2000) investigate the influence of land tenure on the adoption of conservation practices in US. They find that the owner-operators are more likely to adopt practices that provide benefits over the longer term. Furthermore, Gebremedhin and Swinton (2003) examine the role of land tenure security in investing in soil conservation in northern Ethiopia. They find that adoption of technologies associated with long-term investment such as stone terraces was positively influenced by land tenure security. These findings confirm the importance of designing tenancy arrangements that guarantee rights to enjoy long-term benefits derived from implementing technological innovations.

In addition, results show that farmers participating in agrarian organizations are more likely to use irrigation. Participation in organization may serve as a vehicle for the adaptation, likely because transmission and diffusion of knowledge on new technologies are fostered in organizations. These results are in line with Isham (2002). The author finds that households with ethnically based and participatory social affiliations in Tanzania may be more likely to diffuse new information successfully and therefore to adopt new technologies.

We also explore the interactions between some institutional aspects and household characteristics in our adoption models (see Table 8). In particular, we interact age, education, and farm size with the institutional variables secure tenure, extension, and credit. We find that extension services are more likely to promote adoption of irrigation among smaller farmers. This is important for focalizing extension resources. When studying the switch from non-irrigation to modern irrigation, extension services also play a role here, and access to credit seems to help less educated and small farmers more. The latter suggests that institutional aspects are more important for less favored farmers (small and less educated farmers). We do not find significant results for the interacted variables when studying the switch from traditional to modern irrigation.

There are some discrepancies in the results concerning the variables (1) residence in the farm, (2) soil quality, and (3) production risk measures. Regarding (1), results show that farmers who reside at the farm premises are more likely to adopt irrigation technology. In particular, farmers are more willing to shift from non-irrigation to the new technology, as shown by the multinomial probit, and change from traditional irrigation to new technology, as suggested by the probit model with sample selection. These results may be consistent with information diffusion arguments (Foster and Rosenzweig 1995) Bandiera and Rasul 2006), as interactions and networks associated with farming activities are more likely to happen on-farm.

Table 8. Estimates of the models of technology adoption with interactions

Variables	Probit Irrigation = 1	Ordered probit+ Level: 1, 2, 3	Multinomial probit		Probit on selected sample Modern irrigation = 1	Probit with sample selection Modern irrigation = 1
			Modern	Traditional		
Moments of production distribution						
Mean	0.0392 (0.0284)	0.0196 (0.0206)	0.0770 (0.0826)	0.0519 (0.0425)	0.155 (0.0919)*	0.172 (0.0841)**
St. Dev.	0.552 (0.130)***	0.170 (0.0968)*	−0.0352 (0.286)	1.000 (0.205)***	−0.630 (0.338)*	−0.906 (0.287)***
Skewness	0.0436 (0.0160)***	0.00728 (0.0119)	0.00930 (0.0317)	0.0861 (0.0255)***	−0.0458 (0.0404)	−0.0820 (0.0340)**
Household characteristics						
Gender	0.137 (0.0711)*	0.127 (0.0570)**	0.0614 (0.179)	0.220 (0.106)**	−0.347 (0.217)	−0.286 (0.199)
Age	−0.228 (0.299)	−0.0813 (0.237)	0.812 (1.216)	−0.496 (0.441)	2.608 (2.427)	1.985 (1.884)
Education	0.118 (0.0534)**	0.106 (0.0412)**	0.122 (0.186)	0.162 (0.0784)**	−0.0997 (0.286)	−0.117 (0.241)
Reside in farm	0.129 (0.0738)*	0.0612 (0.0564)	0.542 (0.205)***	0.169 (0.109)	0.558 (0.222)**	0.672 (0.207)***
Dependence	0.0207 (0.0312)	−0.00394 (0.0241)	−0.000804 (0.0692)	0.0160 (0.0473)	0.0189 (0.0844)	−0.0155 (0.0766)
Capital	0.00966 (0.00458)**	0.00870 (0.00354)**	0.0269 (0.0111)**	0.00894 (0.00693)	0.0253 (0.0136)*	0.0259 (0.0123)**
Land	0.0757 (0.0950)	0.0241 (0.0714)	0.599 (0.294)**	0.101 (0.139)	0.466 (0.367)	0.520 (0.356)
Institutional aspects						
Secure tenure	0.264 (1.266)	0.788 (1.012)	3.549 (4.951)	−0.228 (1.873)	9.854 (10.35)	8.578 (8.106)
Extension	−0.480 (1.409)	−0.378 (1.130)	3.051 (2.662)	−2.089 (2.252)	1.552 (3.881)	0.129 (3.490)
Credit	0.262 (1.258)	−0.147 (0.951)	−0.901 (2.535)	2.460 (1.954)	−2.204 (3.304)	−1.669 (3.016)
Participation	0.216 (0.0821)***	0.0992 (0.0581)*	0.00871 (0.174)	0.337 (0.123)***	−0.0558 (0.192)	−0.131 (0.178)
Edad × Secure tenure	0.0351 (0.323)	−0.102 (0.256)	−0.830 (1.228)	0.190 (0.478)	−2.524 (2.475)	−2.248 (1.939)
Edad × Extension	0.138 (0.348)	0.135 (0.279)	−0.611 (0.653)	0.477 (0.554)	−0.256 (0.946)	0.0569 (0.852)
Edad × Credit	0.0196 (0.311)	0.117 (0.235)	0.453 (0.626)	−0.561 (0.483)	0.743 (0.807)	0.621 (0.741)
Education × Secure tenure	−0.0403 (0.0566)	−0.0378 (0.0432)	0.102 (0.189)	−0.0862 (0.0834)	0.261 (0.289)	0.279 (0.243)
Education × Extension	0.0160 (0.0588)	0.0131 (0.0449)	−0.0624 (0.102)	0.0473 (0.0953)	−0.00933 (0.132)	0.0487 (0.120)
Education × Credit	−0.0798 (0.0492)	−0.0493 (0.0354)	−0.157 (0.0849)*	−0.149 (0.0812)*	−0.0997 (0.106)	−0.115 (0.0970)
Land × Secure tenure	0.109 (0.0987)	0.109 (0.0737)	0.0212 (0.293)	0.0863 (0.146)	−0.209 (0.378)	−0.353 (0.364)
Land × Extension	−0.191 (0.0809)**	−0.139 (0.0614)**	−0.242 (0.140)*	−0.285 (0.130)**	0.179 (0.163)	0.158 (0.148)
Land × Credit	0.00431 (0.0696)	−0.0401 (0.0510)	−0.230 (0.129)*	0.118 (0.111)	−0.181 (0.144)	−0.186 (0.133)
N° irrigat (*T*)	0.00197 (8.40e−05)***	0.00127 (5.56e−05)***	0.00158 (0.000240)***	0.00302 (0.000120)***	−0.00121 (0.000240)***	−0.00114 (0.000226)***
N° irrigat (*M*)	−0.000829 (0.000501)*	−0.00144 (0.000304)***	0.00317 (0.000889)***	−0.00160 (0.000699)**	0.00310 (0.000825)***	0.00232 (0.000792)***
Environmental factors						
Soil quality	−0.643 (0.139)***	−0.133 (0.110)	−0.872 (0.355)**	−1.004 (0.202)***	0.542 (0.404)	0.530 (0.373)
Rainfall	−1.184 (0.0680)***	−0.886 (0.0511)***	−0.716 (0.181)***	−1.941 (0.0968)***	0.831 (0.173)***	
North zone	1.125 (0.302)***	0.363 (0.123)***	1.777 (0.577)***	1.419 (0.426)***	0.474 (0.353)	−0.373 (0.277)
South zone	−0.559 (0.129)***	−0.894 (0.102)***				
Constant	6.442 (1.289)***		−2.908 (5.131)	11.34 (1.894)***	−17.38 (10.31)*	−9.377 (7.925)
Cut 1		−4.349 (1.019)***				
Cut 2		−1.954 (1.019)*				
Observations	7274	7274	7274		1223	1223

Bootstrapped standard errors in parentheses, 1000 replications

*** p < 0.01, ** p < 0.05, * p < 0.1

Moreover, results from the multinomial probit model show that both traditional and modern irrigation are promoted in zones with poorer soil conditions, probably as a means to deal with these disadvantageous characteristics. However, soil quality considerations do not seem to drive shifts from traditional to modern irrigation. Results are in line with previous studies finding that modern irrigation technologies are more likely used in locations with relatively low land quality (Caswell and Zilberman 1986; Dinar and Yaron 1990; Green et al. 1996).

Regarding production risk, results are twofold. On the one hand, production risk positively affects the shift from non-irrigation to adoption of traditional technology. In other words, our findings suggest that the greater the variance of production, the higher the probability farmers will implement traditional irrigation systems. Similarly, a greater value in the skewness of production entails a higher probability of implementing traditional irrigation. This last finding tells us that farmers care about downside yield risk, assigning negative weights to production distribution that show higher probabilities of obtaining lower production values. On the other hand, adoption of modern irrigation responds negatively to increases in both variance and skewness of production when having traditional irrigation as baseline. Results are robust to regional differences (see Appendix 2).[20] The latter entails that farmers are less willing to shift to modern irrigation when experiencing less certain yields and higher downside risks. While results regarding traditional irrigation support the well-known decreasing-risk properties of irrigation, findings on modern irrigation are not in line with the ability of new technology to hedge against production risk highlighted in the literature before (Koundouri et al. 2006; Torkamani and Shajari 2008).

To verify if these findings can be applicable to larger segment of the Chile's agricultural sector, we conduct a series of robustness checks. First, we replicate our results using information on wheat small-scale farmers. Wheat is one of the most important crops in Chile, comprising more than 50 % of the total hectares devoted to annual crops, and grown in the whole country in diverse climate and soil conditions. To allow enough information in all the irrigation categories, we, here, were less restrictive than in the potato case. We selected those farmers that reported to devote more than 90 % of their hectares to wheat farming. This is an issue in the sense that we cannot guarantee that the inputs observed in the data are indeed used in the wheat production process as in the potato case. This is crucial for estimating production risk measures such that results must be interpreted with caution. Results are shown in Appendix 3. We find similar evidence as in the case of potato farmers. Adoption of traditional irrigation is more likely to occur among farmers with high production risk. However, the risk-reducing effect of modern irrigation technology remains unclear. Second, to gain efficiency, we pool observations of farmers allocating 100 % of land to either potato or wheat. Although the pooling exercise seems to be reasonable, issues regarding the association between inputs and outputs can emerge again when potatoes and wheat are not necessarily grown using the same inputs (or we do not observe that). Results are shown in Appendix 4. We find support for previous findings. Traditional irrigation seems to be risk-decreasing from the perspective of non-irrigators while the switch from traditional to modern irrigation increases production risk. Third, potato and wheat farming are more resilient and less water demanding than other crops such that results regarding the switch from no-irrigation to traditional irrigation (and not modern irrigation) reducing production risk may respond to this specific characteristic. To explore this further, we replicate the results for rice farming, a water-intensive crop. However, exactly the same issue emerges as in other similar crops. The rice sector in Chile is small, and it is not spread out the whole territory, which reduces climate variation. The Census reports 1534 rice farmers mainly located in the central–south regions in Chile. The majority of them use traditional irrigation (96 %) and only a small fraction (4 %) modern irrigation. Farmers without irrigation methods amount to 1.56 %. Given the structure of irrigation adoption in rice farming, it is not possible to control for pre-conditions applying sample selection models such that decisions narrow down to a switch from traditional to modern irrigation. One caveat is that we cannot ensure that irrigation is associated with rice farming since many of the rice farmers grow jointly other crops. When applying the criterion of 100 % land allocation in rice to ensure association between inputs and rice output, we end up with very few observations (86), of which only one farmer reports to use modern irrigation. The low adoption of modern irrigation among more water intensive crops suggests that the insufficient technification in irrigation is a general problem in the agricultural sector in Chile. Having those data issues in mind, we replicate the results for those farmers that report to grow rice (see Appendix 5).

We do not find significant results for the second and/or third moments of the production distribution, although coefficients are negative.

Lack of support of the risk-reducing effect of modern technology may imply that farmers in Chile are facing problems regarding seeing the virtues of water-saving technologies as risk-decreasing inputs, probably due to the higher uncertainty observed in the earlier stages of diffusion. The optimal use of methods of modern irrigation may depend on certain characteristics of soil and climate that are specific to a particular farm and influence the quantities of water required to achieve efficiency in water application. The structure of this relationship may be more difficult to determine in earlier stages. In fact, new irrigation technology in Chile is indeed in an early phase of the technological cycle, where lower levels of knowledge and diffusion may possibly be adding an additional uncertain component that would offset the advantages of modern irrigation. In contrast, when analyzing the adoption of traditional irrigation, we are looking at shifts from a natural state of non-irrigation toward the use of water input. In particular, this technological change seems to be more developed in Chile and has likely reached a phase of maturity in the technological cycle. The latter supports the importance of improving the understanding and modeling of irrigation choices in settings in which rainfed agriculture is still predominant, as typical in developing countries.

Finally, we evaluate the predictive ability of models looking at the percentage of correct predictions.[21] Appendix 6 shows these results. Percentages of correct predictions range from 87 to 96 %, and univariate models explaining modern irrigation choices, having as a baseline traditional irrigation, perform better than multivariate models. Correct predictions as estimating a probit model for modern irrigation choices increases significantly from 1 to 29 %. Correcting for sample selection improves the fitting of the model, adding 14 % extra correct predictions. Furthermore, prediction in all of the remainder irrigation statuses also improves. The latter suggests that modeling non-irrigation as one more level of technology may be inappropriate when there exists specific natural pre-conditions that drive non-irrigation choices. Thus, it seems reasonable to control for these pre-conditions before studying changes from conventional to new irrigation technology.

7 CONCLUSION

New irrigation technology can improve productivity by attaining higher efficiency in water application, which also brings external benefits in the form of water conservation. Furthermore, new irrigation methods hold numerous advantages in zones with limiting environmental conditions, i.e. water scarcity and lower soil quality. In this paper, we study the effect of production risk on the adoption of irrigation. A large fraction of rural households in developing countries does not use either modern or traditional irrigation and is completely exposed to climatic variations. This gives rise to the question on how to include non-irrigators when estimating irrigation choices. We discuss results from an ordered, unordered, and sample selection models.

The results indicate that farmers with a higher educational level, larger capital, credit access, receiving extension services, and those who reside in the plot and in communes with larger number of adopters are more likely to adopt modern irrigation. On the other hand, results for production risk depend on the type of irrigation technology. Whereas production risk promotes adoption of traditional irrigation, it seems to be an obstacle when assessing shifts from traditional to modern irrigation. Credit access, extension services and production variance seem to be more important in terms of magnitude.

However, a couple of caveats deserve attention. First, given the cross-sectional nature of our data, we are not able to fully explore dynamics in irrigation decisions. These may be important, especially in examining the role of economic and policy changes on adoption of modern irrigation. Second, our results may be crop-specific since water needs are supposed to vary across different cropping activities. For example, potato and wheat are relatively more resistant to water stress than other crops such as rice and fruits and therefore a switch from conventional to modern irrigation methods may be more likely to be risk-decreasing in these last cases. Even though we did find sufficient evidence for our sample of rice farmers, data issues do not allow us to state plausibly whether our results can be extrapolated to other relatively more water-demanding cropping activities. Further research examining how results vary across different crops is needed.

Despite these considerations, our findings have important implications for the success of government interventions to address concerns of the low rate of adoption of new irrigation technology. First, adoption of irrigation indeed reduces risk; however, we found that farmers may experience an increase in production risk when switching from traditional to modern irrigation methods. This increased risk may be because of lack of diffusion and knowledge on water-saving technology, which make future profit flows much more uncertain, especially in early stages of the technological cycle. The latter prevents the full exploitation of the advantages of modern irrigation in terms of high productivity and lower production volatility. Our results suggest that policy interventions through extension programs providing adoption-related information and easing access to credit are necessary to induce faster adoption and diffusion among small-scale farmers. Second, we found evidence in favor of sample selection correction as analyzing irrigation choices. Therefore, controlling for pre-conditions that determine irrigation before analyzing changes in technology is crucial to achieve a better understanding on irrigation choices in developing countries where non-irrigators still remain as a significant part of the rural population. Finally, another important implication of our empirical findings is that climate change, which is expected to increase temperature and reduce precipitations in Chile, is likely to have a material effect on the agricultural sector. More specifically, it may increase drought risk and therefore reduce crop yields. Thereby, adoption of more efficient irrigation technologies is crucial to manage the higher climate risk expected for the coming years.

FOOTNOTES

*1*Too much water poured on the ground by traditional irrigation systems may provoke a saturation process, making excess water flow over the land along with pollutants derived from the remains of pesticides and fertilizers.

*2*In the context of locations suffering from water scarcity, conflicting water use and insufficient levels of investment in water supply infrastructure, the adoption of water-saving technologies arises as one alternative to increase efficiency in water use and minimize contamination problems. Furthermore, water-saving technologies contributes to enhance productivity and quality of production through a more uniform and precise irrigation (Dinar and Zilberman 1991; Khana and Zilberman 1997; Zilberman et al. 1997).

*3*The determinants of modern irrigation have been examined by Caswell and Zilberman (1985), Negri and Brooks (1990), Shrestha and Gopalakrishnan (1993), Green et al. (1996), Skaggs (2001), Foltz (2003), Daberkow and Mcbride (2003), Negri et al. (2005), He et al. (2007), Schuck et al. (2005).

*4*First, climatic risk may trigger harvest failure as a result of occurrences of unexpected events from nature such as drought, flood or frost. Second, the inherent volatility of the agricultural market explained mainly by fluctuations in demand and supply generates another source of income variation via price uncertainty. Third, given the fact that farmers make most decisions in advance far before the final product is sold in the market, prices are not known at the time choices regarding the production process are made. Fourth, the recurrent need that governments face to change the direction of agricultural policy produces a further risk component that enhances the uncertainty in investment decisions. Fifth, individual shock may arise, for instance, when households are hit by an illness or unexpected death of one of its members whose contribution to the family budget is vital.

*5*Alternatively, poor farmers get involved in informal risk–coping mechanisms based on agreements made by members of the same group or community to support each other in the case of shocks. In addition, farmers also cope with risk by generating incomes from off-farm activities or by receiving remittances from relatives that are in a better position

*6*The Agricultural and Forestry Census is conducted each 10 years. The previous Census was carried out in 1997. To our knowledge, the 2007 Census was not designed to be able to track a sub-sample of rural households from previous waves. The latter does not allow us to conduct differences in differences techniques while controlling for unobserved time-invariant effects. However, it is highly likely that structural changes affecting differently irrigators and non-irrigators have taken place in such a long period, which would make difference in difference methods somehow problematic.

*7*Eleven farmers report to have both traditional and modern irrigation. Given the reduced number of modern irrigators in our sample, we define them as adopting modern irrigation.

*8*If we assume that the marginal cost of water is closely associated with the energy cost of pumping water, there is a positive relation between depth of water sources and extraction costs.

*9*Due to lack of data on prices, we proxy production risk by the moments of production function rather than the moments of the profit function as Koundouri et al. (2006) do. This assumption lies in the linear relationship between the moments of profit and production function valid under price-taker individuals.

*10*We used a Translog functional form which assumes inputs in levels, squares and cross variables.

*11*1 hectares = 2.5 acres.

*12*The capital variable was built using information with respect to ownership of draft mechanical capital. These were weighted considering market prices. For the construction of the capital variable, we considered the following tools and machinery: Ploughs trucks, vans, carts, choppers, harvesters, cultivators, zero tillage, spray machines, harrows, rakes, reapers, seeders, hoppers, and tractors.

*13*It takes the value of 0 if the farmer has no formal education, 1 if s/he has partially completed elementary school, 2 if s/he has completed elementary education, 3 if s/he has partially completed high school, 4 if s/he graduated from high school, 5 if s/he has partially completed a technical program, 6 if s/he has completed a technical program, 7 if s/he has partially completed her/his university education and 8 if s/he graduated from college/university.

*14*It takes the value 3 if the agricultural income represents 75 % or more of household income, 2 when the proportion is between 50 and 75 %, 1 if it is between 25 and 50 % and 0 if this percentage is less than 25 %.

*15*Rainfall data were obtained from agro-climate information provided by the Chilean Institute of Meteorological Information (CIMI 2011) between the agricultural seasons 1999–2000 and 2005–2006. Climate measures per location were obtained by matching locations with the nearest meteorological stations. Unfortunately, information for regions VIII, IX, X and XIV are not available in the agro-climate yearbooks with the same level of detail as the rest of the zones. For these cases, we used information available in climate yearbooks collected from meteorological stations situated at airports.

*16*Information on erosion was used as a proxy for land quality. It was extracted from a recent study conducted by the Center of Information in Natural Resources (CIREN 2010), addressed at determining the current and potential erosion of soils in Chile. The methodology for determining the level of erosion integrates a set of soil, topographic, climatic and biological characteristics. Thus, erosion will be more severe to the extent that soils are more porous and sandier, fields are more sloped and hold less vegetation as well as in those locations where precipitations are more "aggressive".

*17*A regional analysis based on agro-ecological zones with similar underlying natural conditions can be an alternative manner to account for the initial choice of irrigation. Unfortunately, there is not enough information on the irrigation status to reach convergence in the estimations for agro-ecological zones (no non-irrigators in the north region and very few irrigators in the south). The later suggests that the "non-irrigation" status in some regions may lead to issues of colinearity. Alternatively, we replicate all the estimations considering only those regions where each irrigation status is observed. This implied to drop Regions VI, X and XIV from our sample. Yet, the results remain essentially unchanged. Results are shown in Appendix 1.

*18*Due to the inclusion of generated regressors in adoption equations in the form of moments of the production distribution, standard errors were corrected using bootstrapping techniques.

*19*Multicollinearity is always a concern. For instance, education is most likely highly correlated with income variables such as the percentage of income coming from agriculture, capital and land can ease access to credit, etc. However, correlation coefficients show of correlations. We obtained a coefficient of 0.16 for land and credit, 0.22 for land and capital and 0.27 for dependence on agriculture and capital. The later indicates that this is not likely to be a major problem in this data. Results available upon request.

*20*Due to the inexistence of modern irrigation observations in some regions, we were not capable to control for all regional dummies in all the specifications.

*21*A prediction is assumed correct when the estimated probabilities of observing a determined irrigation status surpass 50 %. Predictions for modern irrigators in the probit model with sample selection were computed using conditional probabilities—that is, probability of success conditional on selection. Furthermore, non-irrigation predictions for this model correspond to those computed using estimates of the selection equation.

DECLARATIONS

Acknowledgements

This study was funded by the Research Support Fund, University of Bio-Bio, Project DIUBB 1567172/RS and Project GI 160317/EF. The authors thank valuable comments received at the seminar of the Development Economics Research Group (DERG), Department of Economics and Department of Food and Resource Economics, University of Copenhagen, and Nordic Conference in Development Economics, Gothenburg, Sweden. Special thanks go to Finn Tarp, Benedikte Bjerge and Marcela Jaime for their valuable comments on earlier versions. The article was also improved by helpful comments from anonymous referees and the editor. The usual caveats apply.

REFERENCES

1. Antle JM (1983) Testing for stochastic structure of production: a flexible moment-based approach. J Bus Econ Stat 1(3):192–201
2. Antle JM (1987) Econometric estimation of producers' risk attitudes. Am J Agric Econ 69:509–522
3. Bahduri A (1973) A study in agricultural backwardness under semi-feudalism. Econ J 83:120–137
4. Bandiera O, Rasul I (2006) Social network and technology adoption in Northern Mozambique. Econ J 116(514):869–902
5. Bardhan P, Udry C (1999) Development microeconomics. Oxford University Press

6. Caswell M, Zilberman D (1985) The choices of irrigation technologies in California. Am J Agric Econ 67(2):224–234

7. Caswell M, Zilberman D (1986) The effects of well depth and land quality on the choice of irrigation technology. Am J Agric Econ 68(4):798–811

8. CIMI (2011) Agro-climatic year books. Seasons 1989–1990:2006–2007

9. CIREN (2010) Determinación de la erosión actual y potencial de los suelos de Chile. http://www.ciren.cl

10. Conley T, Udry C (2010) Learning about a new technology: pineapple in Ghana. Am Econ Rev 100(1):35–69

11. Daberkow S, McBride W (2003) Farm and operator characteristics affecting the awareness and adoption of precision agriculture technologies in the US. Precision Agric 4(2):163–177

12. Dercon S (2002) Income risk, coping strategies and safety nets. World Bank Res Obs 17(2):141–166

13. Dessalegn T (2005) Modelling farm irrigation decisions under under rainfall risk in the White–Volta basis of Ghana. University of Bonn, Thesis

14. Dinar A, Yaron D (1990) Influence of quality and scarcity of inputs on the adoption of modern irrigation technologies. West J Agric Econ 15(2):224–233

15. Dinar A, Zilberman D (1991) The economics of resource-conservation, pollution-reduction technology selection. The case of irrigation water. Resour Energy 13:323–348

16. Doss CR (2006) Analyzing technology adoption using microstudies: limitations, challenges, and opportunities for improvement. Agric Econ 34:207–219

17. Evenson RE, Gollin D (2003) Assessing the impact of the green revolution, 1960–2000. Science 300:758–762

18. FAO (2009) Evolución de la agricultura familiar campesina en Chile en el periodo 1997–2007. Food and Agriculture Organization of the United Nations (FAO). Representation in Chile

19. FAO (2010) Gestión del riesgo de sequía y otros eventos climáticos en Chile. Food and Agriculture Organization of the United Nations (FAO). Representation in Chile

20. Feder G, Just RE, Zilberman D (1985) Adoption of agricultural innovations in developing countries: a survey. Econ Dev Cult Change 33(2):255–298

21. Foltz J (2003) The economics of water-conserving technology adoption in Tunisia: an empirical estimation of farmer technology choice. Econ Dev Cult Change 51(2):359–373

22. Foster A, Rosenzweig M (1995) Learning by doing and learning from others: human capital and Technical change in agriculture. J Polit Econ 103(6):1176–1209

23. Foster A, Rosenzweig M (2010) Microeconomics of technology adoption. Ann Rev Econ 2:395–424

24. Foudi S, Erdlenbruch K (2012) The role of irrigation in farmers' risk management strategies in France. Eur Rev Agric Econ 39(3):439–457

25. Gebremedhin B, Swinton S (2003) Investment in soil conservation in northern Ethiopia: the role of land tenure security and public programs. Agric Econ 29(1):69–84

26. Green G, Sunding D, Zilberman D (1996) Explaining irrigation technology choices. A microparameter approach. Am J Agric Econ 78(4):1064–1072

27. He Xue-Fenget, Huhua C, Feng-Min L (2007) Econometric analysis of the determinants of adoption of rainwater harvesting and supplementary irrigation technology (RHSIT) in the semiarid Loess Plateu of China. Agric Water Manag 89:243–250

28. INDAP (2011) Statistical figures. Instituto de Desarrollo Agropecuario, Santiago, Chile.http://portal.indap.cl/content/view/1106/95/. Accessed 25 May 2011

29. INE (2007) VII Censo Nacional Agropecuario y Forestal, Instituto Nacional de Estadística (INE), Santiago, Chile

30. INIA (2006) Manual de producción de papa para la agricultura familiar campesina (A.F.C), Boletín No. 147, Instituto de Investigaciones Agropecuarias

31. Isham J (2002) The effect of social capital on fertilizer adoption: evidence from rural Tanzania. J Afr Econ 11(1):39–60

32. Just RE, Zilberman D, Rausser GC, Yaron D, Tapiero C (1980) A putty-clay approach to the distributional effects of new technology under risk. In: Operations research in agriculture and water resources. North Holland Publishing Company, New Yourk

33. Khana M, Zilberman D (1997) Incentives, precision technology and environmental protection. Ecol Econ 23:25–43

34. Koundouri P, Nauges E, Tzouvelekas V (2006) Technology adoption under production uncertainty: theory and application to irrigation technology. Am J Agric Econ 88(3):657–670

35. Lee D (2005) Agricultural sustainability and technology adoption: issues and policies for developing countries. Am J Agric Econ 87(5):1325–1334

36. Ministerio de Agricultura (MINAGRI) (2006) Chile potencia alimentaria. Documento base para la elaboración de una agenda pro-competividad, Ministerio de Agricultur

37. Ndiritu S, Kassie M, Shiferaw B (2014) Are there systematic gender differences in the adoption of sustainable agricultural intensification practices? Evidence from Kenya. Food Policy 49:117–12

38. Negri HD, Brooks D (1990) Determinants of irrigation technology choice. West J Agric Econ 15(2):213–223

39. Negri HD, Gollehon NR, Aillery MP (2005) The effect of climate variability on US irrigation adoption. Clim Change 69:299–323

40. Ruttan VW (2002) Productivity growth in world agriculture: sources and constraints. J Econ Perspect 16(4):161–184

41. Schuck E, Frasier W, Webb R, Ellingson L, Umberger W (2005) Adoption of more technically efficient irrigation systems as a drought response. Water Resour Develop 21(4):651–662

42. Shrestha RB, Gopalakrishnan C (1993) Adoption and diffusion of drip irrigation technology: an econometric analysis. Econ Dev Cult Change 41(2):407–418

43. Skaggs RK (2001) Predicting drip irrigation use and adoption in a desert region. Agric Water Manag 51:125–142

44. Soule M, Tegene A, Wiebe K (2000) Land tenure and the adoption of conservation practices. Am J Agric Econ 82(4):993–1005

45. Sunding DL, Zilberman D (2001) The agricultural innovation process: research and technology adoption in a changing agricultural sector. Handb Agric Econ Agric Prod 1:207–261

46. Torkamani J, Shajari S (2008) Adoption of new irrigation technology under production risk. Water Resour Manage 22:229–237

47. Udry C (1994) Risk and insurance in a Rural credit market. An empirical investigation in Northern Nigeria. Rev Econ Stud 61:495–526

48. Van de Ven WPMM, Van Pragg BMS (1981) The demand for deductibles in private health insurance: a probit model with sample selection. J Econom 17:229–252

49. Wooldridge J (2010) Econometric analysis of cross section and panel data, 2nd edn. The Mit Press Cambridge, Massachusett

50. Zilberman D, Khanna M, Lipper L (1997) Economics of new technologies for sustainable development. Aust J Agric Resour Econ 41(1):63–80.

INDEX